IEC 61131-3: Programming Industrial Automation Systems

Karl-Heinz John · Michael Tiegelkamp

IEC 61131-3: Programming Industrial Automation Systems

Concepts and Programming Languages,
Requirements for Programming Systems,
Decision-Making Aids

Second Edition

 Springer

Karl-Heinz John
Irrlrinnig 13
91301 Forchheim
Germany
karlheinz.john@gmx

Michael Tiegelkamp
Kurpfalzstr. 34
90602 Pyrbaum
Germany
Michael.Tiegelkamp@gmx.de

This book contains one Trial DVD. **"SIMATIC STEP 7 Professional, Edition 2006 SR5, Trial License"** encompasses: SIMATIC STEP 7 V5.4 SP4, S7-GRAPH V5.3 SP6, S7-SCL V5.3 SP5, S7-PLCSIM V5.4 SP2 and can be used for trial purposes for 14 days.

This Software can only be used with the Microsoft Windows XP Professional Edition SP3 or Microsoft Windows Vista 32 Bit Business SP1/SP2 or Microsoft Windows Vista 32 Bit Ultimate SP1/SP2 operating systems.

Additional information can be found in the Internet at:
http://www.siemens.com/sce/contact
http://www.siemens.com/sce/modules
http://www.siemens.com/sce/tp

This book also contains one Trial CD-ROM: "Open PCS", a system (full version) for programming with IEC 61131.3, running on any standard Windows PC, using the languages: IL, LD, FBD, SFC, ST and CFC; running under Windows Server 2003, Windows XP SP2 or Windows Vista 32bit. PLC simulation SmartPLC is available for simulating the programs on a PC. The dedicated OPC server SmartPLC/OPC is only required, if additional third-party hardware and/or external OPC clients are connected.

Additional information can be found in the Internet at:
http://www.infoteam.de

ISBN 978-3-642-12014-5   e-ISBN 978-3-642-12015-2
DOI 10.1007/978-3-642-12015-2
Springer Heidelberg Dordrecht London New York

Library of Congress Control Number: 2010925149

Cover design: WMXDesign GmbH, Heidelberg

Printed on acid-free paper

Springer is part of Springer Science+Business Media (www.springer.com)

Preface of the 2nd revised edition

IEC 61131 ("IEC 1131" until 1998) has become widely established in recent years as the programming standard in automation industry. Today, a wide range of small to large PLC manufacturers offer programming systems that are based on this standard. Additional standards and recommendations (e.g. for Motion Control) complement IEC 61131 with functionality in response to growing market requirements.

One of the most important advancements is IEC 61499 (formerly IEC 1499). The basic concepts and ideas of this standard are described in a separate chapter (Chapter 9). Its significance in connection with distributed PLC systems is discussed in Section 7.8.

IEC 61131 is now available in a second edition. The numerous changes and supplements to this standard have been incorporated in the 2nd edition of this book.

A comprehensive index at the end of the book facilitates the search for specific topics.

The enclosed DVD and CD contain the complete demo versions of two programming systems (in the latest versions), enabling the reader to immediately implement and consolidate the knowledge gained from this book by practical application.

We would like to thank SIEMENS AG and infoteam Software AG for providing the enclosed software.

Our special thanks go again to Hans-Peter Otto, member of the IEC and DKE standardisation committees for his active support and mutual inspiration.

With our sincere thanks also to all the people who helped to translate and finish this English version: Andrea Thieme, Kay Thomas-Sukrow, Robie O'Brien, Ormond O'Neill and Michael Sperber.

Above all, we want to thank our families, Susanne, Andreas, Tobias and Andrea, Vera, Olaf, Vanessa and Sebastian, for being so understanding and giving us the freedom to write this book.

We are grateful about the great interest in this book and would like to thank our attentive readers for their numerous suggestions, comments and feedback on typographical errors.

Karl-Heinz John   Michael Tiegelkamp

Winter 2009/2010

Contents

1 Introduction

The rapid advances in performance and miniaturisation in microtechnology are constantly opening up new markets for the programmable logic controller (PLC). Specially designed controller hardware or PC-based controllers, extended by hardware and software with real-time capability, now control highly complex automation processes. This has been extended by the new subject of "safety-related controllers", aimed at preventing injury by machines during the production process.

The different types of PLC cover a wide task spectrum - ranging from small network node computers and distributed *compact units* right up to modular, fault-tolerant, high-performance PLCs. They differ in performance characteristics such as processing speed, networking ability or the selection of I/O modules they support.

Throughout this book, the term PLC is used to refer to the technology as a whole, both hardware and software, and not merely to the hardware architecture. The IEC 61131 programming languages can be used for programming classical PLCs, embedded controllers, industrial PCs and even standard PCs, if suitable hardware (e.g. fieldbus board) for connecting sensors and actors is available.

The broad spectrum of capability of the hardware requires corresponding support from suitable programming tools, to allow low-cost, quality-conscious creation of both simple and complex software solutions. Desirable features of programming tools include:

- Simultaneous use of several PLC programming languages
- "Online" modification of programs in the PLC
- Reverse documentation of the programs from the PLC
- Reusability of PLC program blocks
- "Offline" testing and simulation of user programs
- Integrated configuring and commissioning tools
- Quality assurance, project documentation
- Use of systems with open interfaces.

Modern PCs have enabled increasingly efficient *PLC programming tools* to be developed in the last 10 years.

K.-H. John, M. Tiegelkamp, *IEC 61131-3: Programming Industrial Automation Systems*, 2nd ed., DOI 10.1007/978-3-642-12015-2_1,
© Springer-Verlag Berlin Heidelberg 2010

The classical PLC programming methods, such as the Instruction List, Ladder Logic or Control System Function Chart, which have been employed until now, have reached their limits. Users want uniform, manufacturer-independent language concepts, high-level programming languages and development tools similar to those that have already been in existence in the PC world for many years.

With the introduction of the international standard IEC 1131 (meanwhile renamed to IEC 61131), a basis has been created for uniform PLC programming taking advantage of the modern concepts of software technology. The standard is now available in a revised second edition, which has been fully incorporated into this book.

1.1 Subject of the Book

The aim of this book is to give the reader an understandable introduction to the concepts and languages of standard IEC 61131. Simple examples are given to explain the ideas and application of the new PLC programming languages. An extensive example program summarises the results of each section.

The book serves as a helpful guide and introduction for people in training and at work who want to become acquainted with the possibilities of the new standard. It describes the methods of the standard from a manufacturer-independent perspective. Characteristics and specific versions of individual programming systems should be described in the relevant manuals.

Some experience with personal computers and basic knowledge in the field of PLC technology are required. Experienced PLC programmers will also find information here which will ease working with these programming systems. The book makes a point of describing the standard itself and less the relevant versions of programming systems available on the market.

This book is a useful reference work for students and facilitates the systematic learning of the new programming standard.

Readers can also use the enclosed "Buyer's Guide" to evaluate individual PLC programming systems for themselves. See the enclosed CD-ROM.

The formal contents and structure of the IEC standard are presented in a practice-oriented way. Difficult topics are clearly explained within their context, and the interpretation scope as well as extension possibilities of the standard are demonstrated.

This book is intended to give the reader concrete answers to the following questions:

- How do you program in accordance with IEC 61131? What are the essential ideas of the standard and how can they be applied in practice?
- What are the advantages of the new international standard IEC 61131 compared with other microcontroller programming or PC programming?
- What features must contemporary programming systems have in order to be consistent with IEC 61131 and to fulfil this standard?
- What do users need to look for when selecting a PLC programming system; what criteria are decisive for the performance of programming systems?

Chapter 2 presents the three basic building blocks of the standard: **program, function and function block**. An introductory example which includes the most important language elements of the standard and provides an overview of its programming methods gives an initial introduction to the concepts of IEC 61131.

Chapter 3 describes the **common language elements** of the five programming languages as well as the possibilities of data description with the aid of declarations.

The **five programming languages** of IEC 61131 are explained at length and illustrated by an extensive example in **Chapter 4**.

The strength of IEC 61131 is partly due to the uniform description of frequently used elements, the **standard functions** and **standard function blocks**. Their definition and application are described in **Chapter 5**.

After programming, the programs and the data have to be assigned to the features and hardware of the relevant PLC by means of **configuration**. This is to be found in **Chapter 6**.

The PLC market is developing into a technology with very specific requirements. These **special features of programming** for a PLC as well as their implementation using the facilities of IEC 61131 are the subject of **Chapter 7**.

Chapter 8 summarises the most important qualities of the standard from Chapters 2 to 7. The essential advantages of the standard and of consistent programming systems are outlined here for reference.

Chapter 9 introduces the standard **IEC 61499** for distributed automation processes. It is based on IEC 61131-3, but adopts a wider approach to cater for the demands for parallelism and decentralisation imposed by modern automation tasks.

Chapter 10 explains the use of the enclosed CD-ROM. It includes all programming examples in this book, a buyer's guide in tabular form, and executable versions of two IEC programming systems.

The **Appendices** supply further detailed information.

The **Glossary** in **Appendix I** gives a brief explanation of the most important terms used in this book in alphabetical order.

Appendix J contains the bibliography, which gives **references** not only to books but also to specialised papers on the subject of IEC 61131-3.

Appendix K is a general index which can be very helpful for locating keywords.

1.2 The IEC 61131 standard

In several parts, standard IEC 61131 summarises the requirements of PLC systems. These requirements concern the PLC hardware and the programming system.

The standard includes both the common concepts already in use in PLC programming and additional new programming methods.

IEC 61131-3 sees itself as a **guideline for PLC programming**, not as a rigid set of rules. The enormous number of details defined means that programming systems can only be expected to implement part but not all of the standard. PLC manufacturers have to document this amount: if they want to conform to the standard, they have to prove in which parts they do or do not fulfil the standard.

For this purpose, the standard includes 62 **feature tables** with requirements, which the manufacturer has to fill in with comments (e.g. "fulfilled; not implemented; the following parts are fulfilled:...").

The standard provides a **benchmark** which allows both manufacturers and user to assess how closely each programming system keeps to the standard, i.e. complies with IEC 61131-3.

For further proof of compliance, PLCopen (see Section 1.3) defines further tests for compliance levels which can be carried out by independent institutions.

The standard was established by working group SC65B WG7 (originally: SC65A WG6) of the international standardisation organisation **IEC (International Electrotechnical Commission)** which consists of representatives of different PLC manufacturers, software houses and users. This has the advantage that it is accepted as a guideline by most PLC manufacturers. Thus, IEC 61131-3 has made its way to become the only worldwide standard for PLC programming in recent years.

1.2.1 Goals and benefits of the standard

Because of the constantly increasing complexity of PLC systems there is a steady rise in costs for:

- Training of applications programmers
- The creation of increasingly larger programs
- The implementation of more and more complex programming systems.

PLC programming systems are gradually following the mass software market trend of the PC world. Here too, the pressure of costs can above all be reduced by standardisation, synergy, and reusability.

Manufacturers (PLC hardware and software).
Several manufacturers can invest together in the multi-million dollar software required to fulfil the functionality necessary in today's market.

The basic form of a programming system is determined to a large extent by the standard. Basic software such as editors, with the exception of particular parts like code generators or "online" modules, can be shared. Market differentiation results from supplementary elements to the basic package, which are required in specific market segments, as well as from the PLC hardware. Development costs can be substantially reduced by buying ready-made products. The error proneness of newly developed software can be greatly reduced by the use of previously tested software.

The development costs of contemporary programming tools have increased significantly as a result of the required functionality. By buying ready-made software components (offered only by some of the programming system manufacturers) or complete systems the "time to market" can be significantly shortened, which is essential in order to keep pace with the rapid hardware evolution.

Users
Users often work simultaneously with PLC systems from different manufacturers. Up to now this has meant that employees have needed to take several different training courses in programming, whereas with IEC 61131-3-compliant systems training is limited to the finer points of using the individual programming systems and additional special features of the PLCs. This cuts down on the need for system specialists and training personnel, and PLC programmers are more flexible.

The requirements of the standard ease the selection of suitable programming systems because systems that conform to the standard are easily comparable.

Though it is not expected that complete application programs will be able to be exchanged between different *PLC systems* in the foreseeable future, language elements and program structure are nevertheless similar among the different IEC systems. This facilitates porting onto other systems.

1.2.2 History and components

The standard *IEC 61131* represents a combination and continuation of different standards. It refers to 10 other international standards (IEC 50, IEC 559, IEC 617-12, IEC 617-13, IEC 848, ISO/AFNOR, ISO/IEC 646, ISO 8601, ISO 7185, ISO 7498). These include rules about the employed character code, the definition of the nomenclature used or the structure of graphical representations.

Several efforts have been made in the past to establish a standard for PLC programming technology. Standard IEC 61131 is the first standard to receive the necessary international (and industrial) acceptance. The most important precursor documents to IEC 61131 are listed in Table 1.1.

Year	German	international
1977	DIN 40 719-6 (function block diagrams)	IEC 848
1979		Start of the working group for the first IEC 61131 draft
1982	VDI guideline 2880, sheet 4 PLC programming languages	Completion of the first IEC 61131 draft; Splitting into 5 sub-workgroups
1983	DIN 19239 PLC programming	Christensen Report (Allen Bradley) PLC programming languages
1985		First results of the IEC 65 A WG6 TF3

Table 1.1. Important precursors of IEC 61131-3

The international English version is named IEC 61131 followed by a number. Ed. is short for Edition and indicates the relevant issue status.

The *standard* consists of seven parts (status: December 2009). The overview below shows the relevant titles as well as the year of their first publication and their most recent edition in parentheses:

- IEC 61131-1 Ed. 2: General information (2003) [IEC 61131-1]
- IEC 61131-2 Ed. 3.0: Equipment requirements and tests (1994; 2007) [IEC 61131-2]
- IEC 61131-3 Ed. 2.0: Programming languages (1993; 2003); next revision envisaged for 2010 [IEC 61131-3]
- IEC 61131-4 Ed 2.0: User guidelines (1995; 2004) [IEC 61131-4]
- IEC 61131-5 Ed.1.0: Communications (2000) [IEC 61131-5]
- IEC 61131-7 Ed.1.0: Fuzzy control programming (2000) [IEC 61131-7]
- IEC 61131-8 Ed. 2.0: Guidelines for the application and implementation of programming languages for programmable controllers (1997; 2003) [IEC 61131-8].

In addition, **Corrigenda** are published for some of the standards. They include error descriptions in the currently valid edition of the standard and suggest corrections.

Part 1: General information:
Part 1 contains general definitions and typical functional features which distinguish a PLC from other systems. These include standard PLC properties, for example,

the cyclic processing of the application program with a stored image of the input and output values or the division of labour between programming device, PLC and human-machine interface.

Part 2: Equipment requirements and tests:
This part defines the electrical, mechanical and functional demands on the devices as well as corresponding qualification tests. The environmental conditions (temperature, air humidity etc.) and stress classes of the controllers and of the programming devices are listed.

Part 3: Programming languages:
Here the PLC programming languages widely used throughout the world have been co-ordinated into a harmonised and future-oriented version.

The basic software model and programming languages are defined by means of formal definitions, lexical, syntactical and (partially) semantic descriptions, as well as examples.

Part 4: User guidelines
The fourth part is intended as a guide to help the PLC user in all project phases of automation. Practice-oriented information is given on topics ranging from systems analysis and the choice of equipment to maintenance.

Part 5: Communications:
This part is concerned with communication between PLCs from different manufacturers with each other and with other devices.

In co-operation with ISO 9506 (Manufacturing Message Specification; MMS) conformity classes are defined to allow PLCs to communicate, for example, via networks. These cover the functions of device selection, data exchange, alarm processing, access control and network administration.

Part 6: Safety-related PLC:
The standardisation committee is currently working on the first issue of IEC 61131-6 "Safety-related PLC" with the goal of adapting the requirements of safety standard IEC 61508 ("Functional safety of electrical/electronic/programmable electronic safety-related systems") and machine requirement IEC 62061 ("Safety of machinery – Functional safety of safety-related electrical, electronic and programmable electronic control systems") to PLCs.

Part 7: Fuzzy Control Language:
The goal of this part of the standard is to provide manufacturers and users with a common understanding of the integration of fuzzy control applications based on IEC 61131-3 and to facilitate the portability of fuzzy programs between different manufacturers.

Part 8: Guidelines for the application and implementation of programming languages for Programmable Logic Controllers:
This document offers interpretations for questions not answered by the standard. It includes implementation guidelines, instructions for use by the final user as well as assistance in programming.

The standard describes a modern technology and is therefore subject to strong innovation pressure. This explains why further development of the findings of the standard is being carried out at both national and international level.

This book is concerned with Part 3 "Programming Languages", in short IEC 61131-3. It includes the latest modifications and extensions incorporated with Edition 2 in 2003.

1.3 The Organisation PLCopen

PLCopen [PLCopen Europe] is a manufacturer-independent and product-independent international organisation. Many PLC manufacturers, software houses and independent institutions in Europe and overseas are members of the organisation. Coming from different industry sectors, the members are focused on the harmonisation of controller programming and the development of applications and software interfaces in the IEC 61131-3 environment.

In order to reduce the costs in industrial engineering, uniform specifications and implementation guidelines have been devised. These efforts resulted, for example, in standardised libraries for different application fields, the specification of a conformity level for programming languages, and interfaces for an enhanced exchange of software. The PLCopen expert members are organised in technical committees and define these open standards in co-operation with final users.

1.3.1 Aims of PLCopen

PLCopen was founded in 1992, immediately after publication of the standard IEC 61131-3. At that time, the controller market was highly heterogeneous, with a multitude of programming methods for numerous different types of PLCs. Today, IEC 61131-3 has gained worldwide acceptance as a programming standard and serves as the basis for products offered by many software and hardware manufacturers. Owing to PLCopen, it has thus been incorporated in many different machines and other fields of applications.

Today's control system market poses completely new challenges. PLCopen complies with the market's requirements and – as its core activity – has been defining general standards to make automation more efficient. The latest topics of this standardisation work include:

- *Motion Control and safety functions,*
- *XML data exchange format* for standardising basic data of IEC projects in software systems and
- *Benchmarking* projects for devising a detailed benchmark standard .

New requirements in industry and new products will result in more, new automation tasks in the future. It will remain the mission of PLCopen to reach global harmonisation and a standardised understanding.

PLCopen is not another standardisation committee, but rather a group with a common interest wanting to help existing standards to gain international acceptance. Detailed information can be found on the Internet (http://www.plcopen.org).

1.3.2 Committees and fields of activity

PLCopen is divided into several committees, each of which handles a specific field of interest, as shown in Figure 1.1:

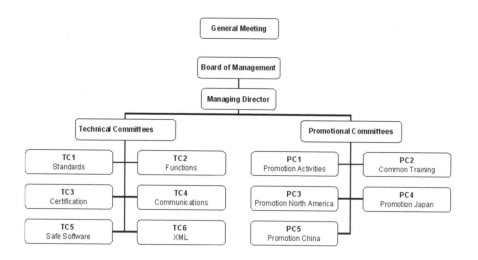

Figure 1.1. Committees of PLCopen

The technical committees work out guidelines for common policy; the promotional committees are responsible for marketing measures.

The work in the committees is carried out exclusively by representatives of individual companies and institutions. This ensures that the resulting papers will be accepted in industry.

1.3.3 Results

As a result of the preparatory work of the promotional committees, PLCopen is represented at several fairs in Europe, USA and the Far East. Workshops and advanced training seminars have brought the desired international recognition for PLCopen.

As a discussion forum for users, manufacturers and software houses some impressive technical results have been achieved:

- *Certification* for manufacturers of PLC programming systems,
- *Exchange format* for user programs.

The committees in detail:

TC 1 – Standards.
This committee is the interface to the international standardisation committees of IEC and OPC [OPC Foundation]. Its members collect suggestions on improvement or error correction for the reponsible IEC 61131 standardisation committee *IEC 65B WG7 working group* [IEC 65B WG7] ; they develop common positions and pass these on to the standardisation committees. In addition, they publish the latest results of the committees' work. In particular, improvements to the 2nd edition of the standard have thus been implemented.

TC 2 – Function and Function Blocks
The members of this committe define libraries of function blocks. The PLCopen Motion Control Specification, for example, has meanwhile become the market standard. This document is subdivided into the following parts:

- Part 1: Basics,
- Part 2: Extensions, additional function blocks,
- Part 3: User guidelines (guidelines and examples for users),
- Part 4: Coordinated motion, focused to the coordinated multi-axes motion in 3D space,
- Part 5: Homing (this function references a specific mechanical position).

TC 3 – Certification.

Several certification levels have been defined and test software has been developed in order to test programming systems for compliance with IEC 61131-3. The test is carried out by institutions accredited by PLCopen to guarantee the desired quality of compliance demonstration. The certification levels include:

- **Base Level (BL):** Original basic definition specifying the basic structure of a program in compliance with the IEC 61131-3 method and the programming manufacturer's declaration of conformity with the standard.
- **Reusability Level (RL):** Functions and function blocks are compatible to an extent that they are portable to different, RL-compliant programming systems.
- **Conformity Level (CL):** Certifies the highest-level of conformance. In many cases, the programming systems utilise only part of the multitude of data types in IEC 61131-3 (26). All data types listed as used in the system by the manufacturer are subjected to a strict conformity test.

The Benchmark working group is also organised under TC 3. It defines specifications and tests for scripts that allow reproducible and portable performance testing of programming systems.

TC 4 – Communication.

TC 4 is involved in definitions at the interface between IEC 61131-3 programming systems and communication systems such as Profibus or CAN.

TC 5 – Safe Software.

This committee makes recommendations for using IEC 61131-3 systems in safety-critical environments. In particular, this includes the new standards IEC 61508 and IEC 61511. Moreover, TC 5 prepares user guidelines covering safety aspects in safety-critical applications.

TC 6 – XML.

TC 6 defines XML schemes for the description of IEC 61131-3 application programs and projects in XML. This includes the textual and graphical languages, variable declaration, and configuration. The specification supports

- the exchange of blocks between systems
- the interface to other software packets such as documentation, simulation, or verification tools.

PC 1 – General Promotion.

This committee is involved in promotional activities in Europe and areas for which there is no separate promotional committee (PCs have also been set up for North America, China, and Japan) [PLCopen Europe].

PC 2 – Common Training.

PC 2 prepares documents for IEC 61131-3 training courses to be held by accredited training institutions [PLCopen Europe].

PC 3 – Promotion North America.
This committee is involved in promotional activities in North America [PLCopen North America].

PC 4 – Promotion Japan.
PC 4 is involved in promotional activities in Japan, in close co-operation with the local PLCopen office [PLCopen Japan].

PC 4 – Promotion China.
PC 5 is involved in promotional activities in China, in close co-operation with the local PLCopen office [PLCopen China].

2 Building Blocks of IEC 61131-3

This chapter explains the meaning and usage of the main language elements of the IEC 61131-3 standard. These are illustrated by several examples from real life, with each example building upon the previous one.

The reader is introduced to the terms and ways of thinking of IEC 61131-3. The basic ideas and concepts are explained clearly and comprehensively without discussing the formal language definitions of the standard itself [IEC 61131-3].

The first section of this chapter gives a compact introduction to the conceptual range of the standard by means of an example containing the most important language elements and providing an overview of the methodology of PLC programming with IEC 61131-3.

The term *"POU" (Program Organisation Unit)* is explained in detail because it is fundamental for a complete understanding of the new language concepts.

As the programming language Instruction List (IL) is already well known to most PLC programmers, it has been chosen as the basis for the examples in this chapter. IL is widespread on the European PLC market and its simple syntax makes it easy to comprehend.

The programming language IL itself is explained in Section 4.1.

2.1 Introduction to the New Standard

IEC 61131-3 not only describes the PLC programming languages themselves, but also offers comprehensive concepts and guidelines for creating PLC projects.

The purpose of this section is to give a short summary of the important terms of the standard without going into details. These terms are illustrated by a simple example. More detailed information will be found in the subsequent sections and chapters.

K.-H. John, M. Tiegelkamp, *IEC 61131-3: Programming Industrial Automation Systems*, 2nd ed., DOI 10.1007/978-3-642-12015-2_2,
© Springer-Verlag Berlin Heidelberg 2010

2.1.1 Structure of the building blocks

POUs correspond to the *Blocks* in previous (conventional) programming systems. POUs can call each other with or without parameters. As the name implies, POUs are the smallest independent software units of a user program.

There are three types of POUs: *Function (FUN)*, *Function block (FB)* and *Program (PROG)*, in ascending order of functionality. The main difference between functions and function blocks is that functions always produce the same result (function value) when called with the same input parameters, i.e. they have no "memory". Function blocks have their own data record and can therefore "remember" status information (*instantiation*). Programs (PROG) represent the "top" of a PLC user program and have the ability to access the I/Os of the PLC and to make them accessible to other POUs.

IEC 61131-3 predefines the calling interface and the behaviour of frequently needed *standard functions (std. FUN)* such as arithmetic or comparison functions, as well as *standard function blocks (std. FB)*, such as timers or counters.

Declaration of variables

The IEC 61131-3 standard uses *variables* to store and process information. Variables correspond to (global) flags or bit memories in conventional PLC systems. However, their storage locations no longer need to be defined manually by the user (as absolute or global addresses), but they are managed automatically by the programming system and each possess a fixed *data type*.

IEC 61131-3 specifies several data types (Bool, Byte, Integer, ...). These differ, for example, in the number of bits or the use of signs. It is also possible for the user to define new data types: user-defined data types such as structures and arrays.

Variables can also be assigned to a certain I/O address and can be battery-backed against power failure.

Variables have different forms. They can be defined (declared) outside a POU and used program-wide, they can be declared as interface parameters of a POU, or they can have a local meaning for a POU. For declaration purposes they are therefore divided into different variable types. All variables used by a POU have to be declared in the declaration part of the POU.

The declaration part of a POU can be written in textual form independently of the programming language used. Parts of the declaration (input and output parameters of the POU) can also be represented graphically.

```
VAR_INPUT                          (* Input variable *)
   ValidFlag      : BOOL;          (* Binary value *)
END_VAR
VAR_OUTPUT                         (* Output variable *)
   RevPM          : REAL;          (* Floating-point value *)
END_VAR
VAR RETAIN                         (* Local variable, battery-backed *)
   MotorNr        : INT;           (* Signed integer *)
   MotorName    : STRING [10];     (* String of length 10 *)
   EmStop AT %IX2.0 : BOOL;        (* Input bit 2.0 of I/O *)
END_VAR
```

Example 2.1. Example of typical variable declarations of a POU

Example 2.1 shows the variable declaration part of a POU. A signed integer variable (16 bits incl. sign) with name MotorNr and a text of length 10 with name MotorName are declared. The binary variable EmStop (emergency stop) is assigned to the I/O signal input 2.0 (using the keyword "AT"). These three variables are known only within the corresponding POU, i.e. they are "local". They can only be read and altered by this POU. During a power failure they retain their value, as is indicated by the qualifier "RETAIN". The value for input variable ValidFlag will be set by the calling POU and have the Boolean values TRUE or FALSE. The output parameter returned by the POU in this example is the floating-point value RevPM.

The Boolean values TRUE and FALSE can also be indicated by "1" and "0".

Code part of a POU

The code part, or instruction part, follows the declaration part and contains the instructions to be processed by the PLC.

A POU is programmed using either the textual programming languages Instruction List (IL) and Structured Text (ST) or the graphical languages Ladder Diagram (LD) and Function Block Diagram (FBD). IL is a programming language closer to machine code, whereas ST is a high-level language. LD is suitable for Boolean (binary) logic operations. FBD can be used for programming both Boolean (binary) and arithmetic operations in graphical representation.

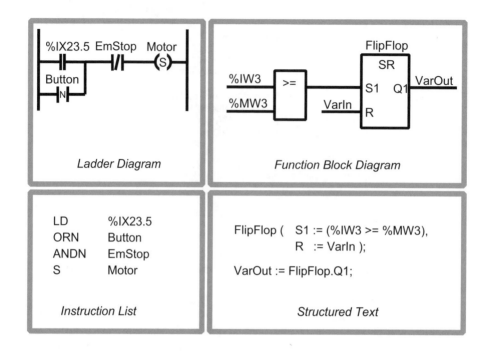

Figure 2.1. Simple examples of the programming languages LD, FBD, IL and ST. The examples in LD and IL are equivalent to one another, as are those in FBD and ST.

Additionally, the description language Sequential Function Chart (SFC) can be used to describe the structure of a PLC program by displaying its sequential and parallel execution. The various subdivisions of the SFC program (steps and transitions) can be programmed independently using any of the IEC 61131-3 programming languages.

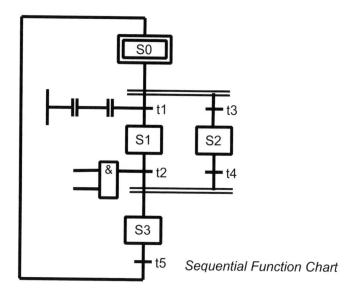

Sequential Function Chart

Figure 2.2. Schematic example of structuring using SFC. The execution parts of the steps (S0 to S3) and the transitions (t1 to t5) can be programmed using any other programming language.

Figure 2.2 shows an SFC example: Steps S0, S1 and S3 are processed sequentially. S2 can be executed alternatively to S1. Transitions t1 to t5 are the conditions which must be fulfilled before proceeding from one step to the next.

2.1.2 Introductory example written in IL

An example of an IEC 61131-3 program is presented in this section. Figure 2.3 shows its POU calling hierarchy in tree form.

This example is not formulated as an executable program, but simply serves to demonstrate POU structuring.

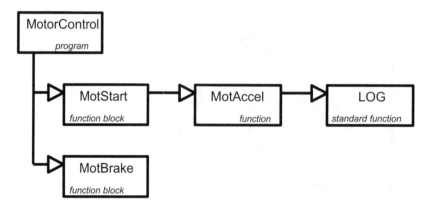

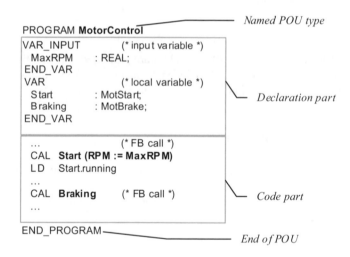

Figure 2.3. Calling hierarchy of POUs in the example

The equivalent IL representation is shown in Example 2.2.

Example 2.2. Declaration of the program MotorControl from Figure 2.3 together with corresponding code parts in IL. Comments are represented by the use of brackets: "(* ... *)".

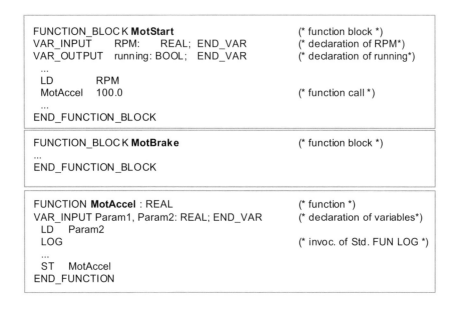

```
FUNCTION_BLOCK MotStart            (* function block *)
VAR_INPUT     RPM:     REAL; END_VAR    (* declaration of RPM*)
VAR_OUTPUT  running: BOOL;  END_VAR     (* declaration of running*)
  ...
  LD       RPM
  MotAccel  100.0                   (* function call *)
  ...
END_FUNCTION_BLOCK
```

```
FUNCTION_BLOCK MotBrake            (* function block *)
  ...
END_FUNCTION_BLOCK
```

```
FUNCTION MotAccel : REAL            (* function *)
VAR_INPUT Param1, Param2: REAL; END_VAR   (* declaration of variables*)
  LD    Param2
  LOG                               (* invoc. of Std. FUN LOG *)
  ...
  ST    MotAccel
END_FUNCTION
```

Example 2.3. The three subprograms of Fig. 2.3 in IL. LOG (logarithm) is predefined standard function of IEC 61131-3.

MotorControl is the main program. When this program is started, the variable RPM is assigned an initial value passed with the call (as will be seen later). This POU then calls the block Start (MotStart). This POU in turn calls the REAL function MotAccel with two input parameters (RPM and 100.0). This then calls LOG – the IEC 61131 *standard function* "Logarithm". After processing Start (MotStart), MotorControl is activated again, evaluates the result running and then calls Braking, (MotBrake).

As shown in Example 2.2, the function blocks MotStart and MotBrake are not called directly using these names, but with the so-called "*instance* names" Start and Braking respectively.

2.1.3 PLC assignment

Each PLC can consist of multiple processing units, such as CPUs or special processors. These are known as *resources* in IEC 61131-3. Several programs can run on one resource. The programs differ in priority or execution type (periodic/cyclic or by interrupt). Each program is associated with a *task*, which makes it into a *run-time program*. Programs may also have multiple associations (*instantiation*).

Before the program described in Examples 2.2 and 2.3 can be loaded into the PLC, more information is required to ensure that the associated task has the desired properties:

- On which PLC type and which resource is the program to run?
- How is the program to be executed and what priority should it have?
- Do variables need to be assigned to physical PLC addresses?
- Are there global or external variable references to other programs to be declared?

This information is stored as the *configuration*, as illustrated textually in Example 2.4 and graphically in Figure 2.4.

```
CONFIGURATION MotorCon
  VAR_GLOBAL Trigger AT %IX2.3 : BOOL;  END_VAR
  RESOURCE Res_1 ON CPU001
    TASK T_1        (INTERVAL := t#80ms,  PRIORITY := 4);
    PROGRAM MotR  WITH T_1 : MotorControl (MaxRPM := 12000);
  END_RESOURCE
  RESOURCE Res_2 ON CPU002
    TASK T_2        (SINGLE := Trigger, PRIORITY := 1);
    PROGRAM MotP  WITH T_2 : MotorProg (...);
  END_RESOURCE
END_CONFIGURATION
```

Example 2.4. Assignment of the programs in Example 2.3 to tasks and resources in a "configuration"

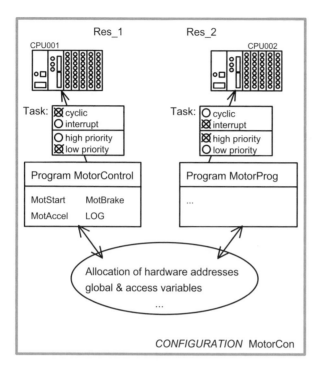

Figure 2.4. Assignment of the programs of a motor control system MotorCon to tasks in the PLC "configuration". The resources (processors) of the PLC system execute the resulting run-time programs.

Figure 2.4 continues Example 2.3. Program MotorControl runs together with its FUNs and FBs on resource CPU001. The associated task specifies that MotorControl should execute cyclically with low priority. Program MotorProg runs here on CPU002, but it could also be executed on CPU001 if this CPU supports multitasking.

The configuration is also used for assigning variables to I/Os and for managing global and communication variables. This is also possible within a PROGRAM.

A PLC project consists of POUs that are either shipped by the PLC manufacturer or created by the user. User programs can be used to build up libraries of tested POUs that can be used again in new projects. IEC 61131-3 supports this aspect of software *re-use* by stipulating that functions and function blocks have to remain "universal", i.e. hardware-independent, as far as possible.

After this short summary, the properties and special features of POUs will now be explained in greater detail in the following sections.

2.2 The Program Organisation Unit (POU)

IEC 61131-3 calls the blocks from which programs and projects are built Program
Organisation Units (POUs). POUs correspond to the program blocks, organisation
blocks, sequence blocks and function blocks of the conventional PLC programming world.

One very important goal of the standard is to restrict the variety and often
implicit meanings of block types and to unify and simplify their usage.

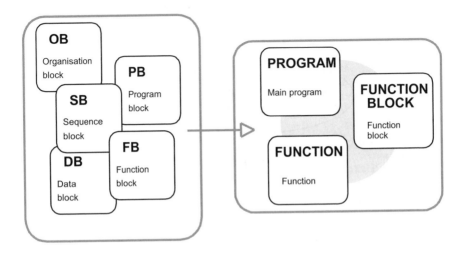

Block types used in DIN 19239   POUs in IEC 61131-3

Figure 2.5. Evolution from previous block types (e.g. German DIN 19239) to the POUs of
IEC 61131-3

As Figure 2.5 shows, IEC 61131-3 reduces the different block types of PLC
manufacturers to three unified basic types. Data blocks are replaced by FB data
memories ("instances", see below) or global multi-element variables (see also
Chapter 3).

The following three *POU types* or "block types" are defined by the new
standard:

POU type	Keyword	Meaning
Program	PROGRAM	Main program including assignment to I/O, global variables and access paths
Function block	FUNCTION_BLOCK	Block with input and output variables; this is the most frequently used POU type
Function	FUNCTION	Block with function value, input and output variables for extension of the basic PLC operation set

Table 2.1. The three POU types of IEC 61131-3 with their meanings

These three POU types differ from each other in certain features:
- **Function (FUN).** POU that can be assigned parameters, but has no static variables (without memory), which, when invoked with the same input parameters, always yields the same result as its function value (output).
- **Function block (FB).** POU that can be assigned parameters and has static variables (with memory). An FB (for example a counter or timer block), when invoked with the same input parameters, will yield values which also depend on the state of its internal (VAR) and external (VAR_EXTERNAL) variables, which are retained from one execution of the function block to the next.
- **Program (PROG).** This type of POU represents the "main program". All variables of the whole program that are assigned to physical addresses (for example PLC inputs and outputs) must be declared in this POU or above it (Resource, Configuration). In all other respects it behaves like an FB.

PROG and FB can have both input and output parameters. Functions, on the other hand, have input and output parameters and a function value as return value. These properties were previously confined to "function blocks".

The IEC 61131-3 FUNCTION_BLOCK with input **and** output parameters roughly corresponds to the conventional function block. The POU types PROGRAM and FUNCTION do not have direct counterparts in blocks as defined in previous standards, e.g. DIN 19239.

A POU is an encapsulated unit, which can be compiled independently of other program parts. However, the compiler needs information about the calling interfaces of the other POUs that are called in the POU ("prototypes"). Compiled POUs can be linked together later in order to create a complete program.

The name of a POU is known throughout the project and may therefore only be used once. Local subroutines as in some other (high-level) languages are not permitted in IEC 61131-3. Thus, after programming a POU (*declaration*), its name and its calling interface will be known to all other POUs in the project, i.e. the POU name is always global.

This independence of POUs facilitates extensive modularisation of automation tasks as well as the *re-use* of already implemented and tested software units.

In the following sections the common properties of the different types of POUs will first be discussed. The POU types will then be characterised, the calling relationships and other properties will be described, and finally the different types will be summarised and compared.

2.3 Elements of a POU

A POU consists of the elements illustrated in Figure 2.6:

- POU type and name (and data type in the case of functions)
- Declaration part with variable declarations
- POU body with instructions.

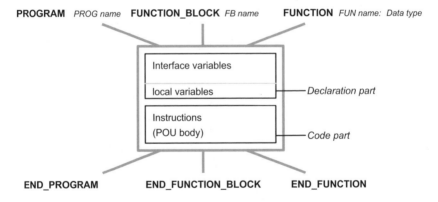

Figure 2.6. The common structure of the three POU types Program (left), Function Block (centre) and Function (right). The declaration part contains interface and local variables.

Declarations define all the variables that are to be used within a POU. Here a distinction is made between variables visible from outside the POU (POU interface) and the local variables of the POU. These possibilities will be explained in the next section and in more detail in Chapter 3.

Within the *code part* (body) of a POU the logical circuit or algorithm is programmed using the desired programming language. The languages of IEC 61131-3 are presented and explained in Chapter 4.

Declarations and instructions can be programmed in graphical or textual form.

2.3.1 Example

The elements of a POU are illustrated in Example 2.5.

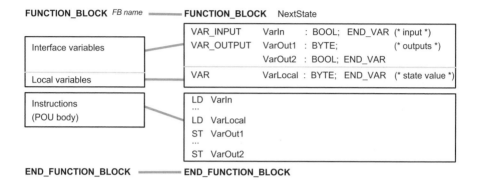

Example 2.5. Elements of a POU (left) and example of a function block in IL (right). The FB contains the input parameter VarIn as well as the two return values VarOut1 and VarOut2. VarLocal is a local variable.

The function block NextState written in IL contains the input parameter VarIn, the two return values VarOut1 and VarOut2 and the local variable VarLocal. In the FB body the IL operators LD (Load) and ST (Store) are used.

FUNCTION_BLOCK NextState

END_FUNCTION_BLOCK

Example 2.6. Graphical representation of the calling interface of FB NextState in Example 2.5.

When using the graphical representation of the calling interface, local FB variables such as VarLocal are not visible.

2.3.2 Declaration part

In IEC 61131-3 variables are used for initialising, processing and storing user data. The variables have to be *declared* at the beginning of each POU, i.e. their assignment to a specific *data type* (such as BYTE or REAL) is made known.

Other *attributes* of the variables, such as battery backup, initial values or assignment to physical addresses, can also be defined during declaration.

As shown by Example 2.7, the declaration of POU variables is divided into separate sections for the different *variable types*. Each *declaration block* (VAR_*...END_VAR) corresponds to one variable type and can contain one or more variables. As Example 2.8 shows, the order and number of blocks of the same variable type can be freely determined or can depend on how the variables are used in a particular programming system.

```
(* Local variable *)
VAR              VarLocal : BOOL;  END_VAR   (* local Boolean variable *)
(* Calling interface: input parameters *)
VAR_INPUT        VarIn : REAL;     END_VAR   (* input variable *)
VAR_IN_OUT       VarInOut : UINT;  END_VAR   (* input and output variable *)
(* Return values: output parameters *)
VAR_OUTPUT       VarOut : INT;     END_VAR   (* output variable *)
(* Global interface: global/external variables and access paths *)
VAR_EXTERNAL  VarGlob : WORD;  END_VAR   (* external from other POU *)
VAR_GLOBAL    VarGlob : WORD;  END_VAR   (* global for other POUs *)
VAR_ACCESS    VarPath : WORD;  END_VAR   (* access path of configuration *)
```

Example 2.7. Examples of the declarations of different variable types

```
(* Declaration block 1 *)
VAR            VarLocal1, VarLocal2, VarLocal3: BOOL; END_VAR
(* Declaration block 2 *)
VAR_INPUT    VarIn1 : REAL;                    END_VAR
(* Declaration block 3 *)
VAR_OUTPUT VarOut : INT;                       END_VAR
(* Declaration block 4 *)
VAR            VarLocal4, VarLocal5 : BOOL;    END_VAR
(* Declaration block 5 *)
VAR_INPUT    VarIn2, VarIn3 : REAL;            END_VAR
(* Declaration block 6 *)
VAR_INPUT    VarIn4 : REAL;                    END_VAR
```

Example 2.8. Examples of declaration blocks: the order and number of the blocks is not specified in IEC 61131-3.

Types of variables in POUs.

As shown by Table 2.2, different types of variables may be used depending on the POU type:

Variable type	Permitted in:		
	PROGRAM	*FUNCTION_BLOCK*	*FUNCTION*
VAR	yes	yes	yes
VAR_INPUT	yes	yes	yes
VAR_OUTPUT	yes	yes	yes
VAR_IN_OUT	yes	yes	yes
VAR_EXTERNAL	yes	yes	no
VAR_GLOBAL	yes	no	no
VAR_ACCESS	yes	no	no
VAR_TEMP	yes	yes	no

Table 2.2. Variable types used in the three types of POU

As Table 2.2 shows, all variable types can be used in programs. Function blocks cannot make global variables available to other POUs. This is only permitted in programs, resources and configurations. FBs access global data using the variable type VAR_EXTERNAL.

Functions have the most restrictions because only local and input and output variables are permitted in them. They return their calculation result using the function return value.

Except for local variables, all variable types can be used to import data into and export data from a POU. This makes data exchange between POUs possible. The features of this *POU interface* will be considered in more detail in the next section.

Characteristics of the POU interface

The POU interfaces, as well as the local data area used in the POU, are defined by means of assigning POU variables to variable types in the declaration blocks. The POU interface can be divided into the following sections:

- Calling or invocation interface: formal parameters (input and input/output parameters)
- Return values: output parameters or function return values
- Global interface with global/external variables and access paths.

The calling interface and the return values of a POU can also be represented graphically in the languages LD and FBD.

The variables of the calling interface are also called *formal parameters*. When calling a POU the formal parameters are replaced with *actual parameters*, i.e. assigned actual (variable) values or constants.

In Example 2.3 FB MotStart has only one formal parameter RPM, which is given the value of the actual parameter MaxRPM in Example 2.2, and it also has the output parameter running. The function MotAccel has two formal parameters (one of which is assigned the constant 100.0) and returns its result as the function return value

This is summarised by Table 2.3.

	Variable types	Remarks
Calling interface (formal parameters)	VAR_INPUT, VAR_IN_OUT	Input parameters, can also be graphically displayed
Return values	VAR_OUTPUT	Output parameters, can also be graphically displayed
Global interface	VAR_GLOBAL, VAR_EXTERNAL, VAR_ACCESS	Global data
Local values	VAR, VAR_TEMP	POU internal data

Table 2.3. Variable types for interface and local data of a POU. See the comments in Example 2.7.

Formal input parameter (VAR_INPUT): Actual parameters are passed to the POU as *values*, i.e. the variable itself is not used, but only a copy of it. This ensures that this input variable cannot be changed within the called POU. This concept is also known as "call-by-value".

Formal input/output parameter (VAR_IN_OUT): Actual parameters are passed to the called POU in the form of a pointer to their storage location, i.e. the variable itself is used. It can thus be read and changed by the called POU. Such changes have an automatic effect on the variables declared outside the called POU. This concept is also known as "call-by-reference".

By working with references to storage locations this variable type provides pointers like those used in high-level languages like C for return values from subroutines.

Formal output parameters, return values (VAR_OUTPUT) are not passed to the called POU, but are provided by that POU as values. They are therefore not part of the calling interface. They appear together with VAR_INPUT and VAR_IN_OUT in graphical representation, but in textual languages such as IL or ST their **values** are read **after** calling the POU.

The method of passing to the calling POU is also "return-by-value", allowing the values to be read by the calling instance (FB or PROG). This ensures that the output parameters of a POU are protected against changes by a calling POU. When a POU of type PROGRAM is called, the output parameters are provided together with the actual parameters by the resource and assigned to appropriate variables for further processing (see examples in Chapter 6).

If a POU call uses complex arrays or data structures as variables, the use of VAR_IN_OUT results in more efficient programs, as it is not the variables themselves (VAR_INPUT and VAR_OUTPUT) that have to be copied at run time, but only their respective pointers. However such variables are not protected against (unwelcome) manipulation by the called POU.

External and internal access to POU variables

Formal parameters and return values have the special property of being **visible** outside their POU: the calling POU can (but need not) use their names explicitly for setting input values.

This makes it easier to document the POU calling interface and parameters may be omitted and/or their sequence may be altered. In this context input and output variables are also protected against unauthorised reading and writing.

Table 2.4 summarises all variable types and their meaning. Access rights are given for each variable type, indicating whether the variable:

- is visible to the calling POU ("external") and can be read or written to there
- can be read or written to within the POU ("internal") in which it is defined.

Variable type	Access rights[a]		Explanation
	external	internal	
VAR, VAR_TEMP *Local Variables*	-	RW	A local variable is only visible within its POU and can be processed only there.
VAR_INPUT *Input Variables*	W[b]	R	An input variable is visible to the calling POU and may be written to (changed) there. It may not be changed within its own POU.
VAR_OUTPUT *Output Variables*	R	RW	An output variable is visible to the calling POU and may only be read there. It can be read and written to within its own POU.
VAR_IN_OUT *Input and Output Variables*	RW	RW	An input/output variable possesses the combined features of VAR_INPUT and VAR_OUTPUT: it is visible and may be read or changed within or outside its POU.
VAR_EXTERNAL *External Variables*	RW	RW	An external variable is required to enable read and write access from within a POU to a global variable of another POU. It is visible only to POUs that list this global variable under VAR_EXTERNAL; all other POUs have no access to this global variable. Identifier and type of a variable under VAR_EXTERNAL must coincide with the corresponding VAR_GLOBAL declaration in PROGRAM.
VAR_GLOBAL *Global Variables*	RW	RW	A variable declared as GLOBAL may be read and written to by several POUs. For this purpose, the variable must be listed with identical name and type under VAR-EXTERNAL in the other POUs.
VAR_ACCESS *Access Paths*	RW	RW	Global variable of configurations as communication channel between components (resources) of configurations (see also Chapter 6). It can be used like a global variable within a POU.

a   W=Write, R=Read, RW=Read and Write
b   can be written to only as formal parameter during invocation

Table 2.4. The meaning of the variable types. The left-hand column contains the keyword of each variable type in bold letters. In the "Access rights" column the read/write rights are indicated for the calling POU (external) and within the POU (internal) respectively.

IEC 61131-3 provides extensive access protection for input and output variables, as shown in Table 2.4 for VAR_INPUT and VAR_OUTPUT: input variables may not be changed within their POU, and output parameters may not be changed outside.

The special relevance of the declaration part to function blocks will be discussed again in Section 2.4.1 when explaining FB instantiation.

The following examples show both external (calling the POU) and internal (within the POU) access to formal parameters and return values of POUs:

```
FUNCTION_BLOCK FBTwo                FUNCTION_BLOCK FBOne
VAR_INPUT                           VAR ExampleFB : FBTwo; END_VAR
 VarIn    : BYTE;
END_VAR
VAR_OUTPUT
 VarOut   : BYTE;
END_VAR
VAR
 VarLocal : BYTE;
END_VAR
...                                 ...
LD    VarIn                         LD    44
AND   VarLocal                      ST    ExampleFB.VarIn
ST    VarOut                        CAL   ExampleFB       (* FB call *)
...                                 LD    ExampleFB.VarOut
                                    ...
END_FUNCTION_BLOCK                  END_FUNCTION_BLOCK
```

Example 2.9. Internal access (on the left) and external access (on the right) to the formal parameters VarIn and VarOut.

In Example 2.9 FBOne calls block ExampleFB (described by FBTwo). The input variable VarIn is assigned the constant 44 as actual parameter, i.e. this input variable is visible and initialised in FBOne. VarOut is also visible here and can be read by FBOne. Within FBTwo VarIn can be read (e.g. by LD) and VarOut can be written to (e.g. using the instruction ST).

Further features and specialities of variables and variable types will be explained in Section 3.4.

2.3.3 Code part

The instruction or code part (body) of a POU immediately follows the declaration part and contains the instructions to be executed by the PLC. IEC 61131-3 provides five programming languages (three of which have graphical representation) for application-oriented formulation of the control task.

As the method of programming differs strongly between these languages, they are suitable for different control tasks and application areas.

Here is a general guide to the languages:

SFC	*Sequential Function Chart*: For breaking down the control task into parts which can be executed sequentially and in parallel, as well as for controlling their overall execution. SFC very clearly describes the program flow by defining which actions of the controlled process will be enabled, disabled or terminated at any one time. IEC 61131-3 emphasises the importance of SFC as an *Aid for Structuring* PLC programs.
LD	*Ladder Diagram*: Graphical connection ("circuit diagram") of Boolean variables (contacts and coils), geometrical view of a circuit similar to earlier relay controls. POUs written in LD are divided into sections known as *networks*.
FBD	*Function Block Diagram*: Graphical connection of arithmetic, Boolean or other functional elements and function blocks. POUs written in FBD are divided into *networks* like those in LD. *Boolean* FBD networks can often be represented in LD and vice versa.
IL	*Instruction List*: Low-level machine-oriented language offered by most of the programming systems
ST	*Structured Text*: High-level language (similar to PASCAL) for control tasks as well as complex (mathematical) calculations.

Table 2.5. Features of the programming languages of IEC 61131-3

In addition, the standard explicitly allows the use of other programming languages (e.g. C or BASIC), which fulfil the following basic requirements of PLC programming:

- The use of variables must be implemented in the same way as in the other programming languages of IEC 61131-3, i.e. compliant with the declaration part of a POU.
- Calls of functions and function blocks must adhere to the standard, especially calls of standard functions and standard function blocks.
- There must be no inconsistencies with the other programming languages or with the structuring aid SFC.

Details of these standard programming languages, their individual usage and their representation are given in Chapter 4.

2.4 The Function Block

Function blocks are the main building blocks for structuring PLC programs. They are called by programs and FBs and can themselves call functions as well as other FBs.

In this section the basic features of function blocks will be explained. A detailed example of an FB can be found in Appendix C.

The concept of the "instantiation of FBs" is of great importance in IEC 61131-3 and is an essential distinguishing criterion between the three POU types. This concept will therefore be introduced before explaining the other features of POUs.

2.4.1 Instances of function blocks

What is an "instance"?

The creation of variables by the programmer by specifying the variable's name and data type in the declaration is called *instantiation*.

In the following Example 2.10 the variable Valve is an *instance* of data type BOOL:

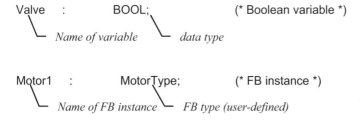

Example 2.10. Declaration of a variable as an "instance of a data type" (top). Declaration of an FB "variable" as an "instance of a user-defined FB type" (bottom).

Function blocks also are instantiated like variables: In Example 2.10 the FB instance Motor1 is declared as an instance of the user-defined function block (FB type) MotorType in the declaration part of a POU. After instantiation an FB can be used (as an instance) and called within the POU in which it is declared.

This principle of instantiation may appear unusual at first sight but, in fact, it is nothing new.

Up until now, for example, function blocks for counting or timing, known for short as counters and timers respectively, were mostly defined by their type (such as direction of counting or timing behaviour) and by a number given by the user, e.g. Counter "C19".

Instead of this absolute number the standard IEC 61131-3 requires a (symbolic) variable name combined with the specification of the desired timer or counter type. This has to be declared in the declaration part of the POU. The programming system can automatically generate internal, absolute numbers for these FB variables when compiling the POU into machine code for the PLC.

With the aid of these variable names the PLC programmer can use different timers or counters of the same type in a transparent manner and without the need to check name conflicts.

By means of instantiation IEC 61131-3 unifies the usage of manufacturer-dependent FBs (typically timers and counters) and user-defined FBs. Instance names correspond to the symbolic names or so-called *symbols* used by many PLC programming systems. Similarly, an FB type corresponds to its calling interface. In fact, FB instances provide much more than this: "Structure" and "Memory" for FBs will be explained in the next two subsections.

The term "function block" is often used with two slightly different meanings: it serves as a synonym for the FB instance name as well as for the FB type (= name of the FB itself). In this book "function block" means *FB type,* while an FB instance will always be explicitly indicated as an instance name.

Example 2.11 shows a comparison between the declarations of function blocks (here only standard FBs) and variables:

```
VAR
    FillLevel      : UINT;    (* unsigned integer variable *)
    EmStop         : BOOL;    (* Boolean variable *)
    Time9          : TON;     (* timer of type on-delay *)
    Time13         : TON;     (* timer of type on-delay *)
    CountDown      : CTD;     (* down-counter *)
    GenCounter     : CTUD;    (* up-down counter *)
END_VAR
```

Example 2.11. Examples of variable declaration and instantiation of standard function blocks (bold).

Although in this example Time9 and Time13 are based on the same FB type (TON) of a timer FB (on-delay), they are independent timer blocks which can be separately called as instances and are treated independently of each other, i.e. they represent two different "timers".

FB instances are visible and can be used within the POU in which they are declared. If they are declared as global, all other POUs can use them as well (with VAR_EXTERNAL).

Functions, on the other hand, are always visible project-wide and can be called from any POU without any further need of declaration. Similarly FB **types** are known project-wide and can be used in any POU for the declaration of instances.

The declaration and calling of standard FBs will be described in detail in Chapter 5. Their usage in the different programming languages is explained in Chapter 4.

Instance means "structure".
The concept of instantiation, as applied in the examples of timer or counter FBs, results in **structured** variables, which:

- describe the FB calling interface like a data structure,
- contain the actual status of a timer or counter,
- represent a method for calling FBs.

This allows flexible parameter assignment when calling an FB, as can be seen below in the example of an up/down counter:

```
VAR
  Counter :    CTUD;       (* up/down counter *)
END_VAR
```

Example 2.12. Declaration of an up/down counter with IEC 61131-3

After this declaration the inputs and outputs of this counter can be accessed using a data structure defined implicitly by IEC 61131-3. In order to clarify this structure Example 2.13 shows it in an alternative representation.

```
TYPE CTUD :      (* data structure of an FB instance of FB type CTUD *)
  STRUCT
    (* inputs *)
    CU :          BOOL;    (* count up *)
    CD :          BOOL;    (* count down *)
    R  :          BOOL;    (* reset *)
    LD :          BOOL;    (* load *)
    PV :          INT;     (* preset value *)
    (* outputs *)
    QU :          BOOL;    (* output up *)
    QD :          BOOL;    (* output down *)
    CV :          INT;     (* current value *)
  END_STRUCT;
END_TYPE
```

Example 2.13. Alternative representation of the data structure of the up/down counter (standard FB) in Example 2.12

The data structure in Example 2.13 shows the formal parameters (calling interface) and return values of the standard FB CTUD. It represents the caller's view of the FB. Local or external variables of the POU are kept hidden.

This data structure is managed automatically by the programming or run-time system and is easy to use for assigning parameters to FBs, as shown in Example 2.14 in the programming language IL:

```
LD     34
ST     Counter.PV      (* preset count value *)
LD     %IX7.1
ST     Counter.CU      (* count up *)
LD     %M3.4
ST     Counter.R       (* reset counter *)
CAL    Counter         (* invocation of FB with actual parameters *)
LD     Counter.CV      (* get current count value *)
```

Example 2.14. Parameterisation and invocation of the up/down counter in Example 2.12

In this example the instance Counter is assigned the parameters 34, %IX7.1 and %M3.4, before Counter is called by means of the instruction CAL (shown here in bold type). The current counter value can then be read.

As seen in Example 2.14, the inputs and outputs of the FB are accessed using the FB instance name and a separating period. This procedure is also used for structure elements (see Section 3.5.2).

Unused input or output parameters are given initial values that can be defined within the FB itself.

In Section. 4.1.4 further methods of calling FBs in IL by means of their instance names are shown.

Instance means "memory".

When several variable names are declared for the same FB type a sort of "FB data copy" is created for each instance in the PLC memory. These copies contain the values of the local (VAR) and the input or output variables (VAR_INPUT, VAR_OUTPUT), but not the values for VAR_IN_OUT (these are only pointers to variables, not the values themselves) or VAR_EXTERNAL (these are global variables).

This means that the instance can store local data values and input and output parameters over several invocations, i.e. it has a kind of "memory". Such a memory is important for FBs such as flip-flops or counters, as their behaviour is dependent on the current **status** of their flags and counter values respectively.

All variables of this memory are stored in a memory area which is **firmly** assigned to this one FB instance (by declaration). This memory area must therefore be **static**. This also means that the stack cannot be used in the usual way to manage local **temporary** variables!

Particularly in the case of function blocks which handle large data areas such as tables or arrays, this can lead to (unnecessarily) large static memory requirements for FB instances.

IEC 61131-3 has therefore defined variable type VAR_TEMP. A value of a variable that does **not** have to be maintained between calls is defined with the VAR_TEMP declaration. In this case the programming system uses a dynamic area or stack to create memory space that is valid only while the instance is executed.

Furthermore, large numbers of input and output parameters can lead to memory-consuming FB instances. The use of VAR_IN_OUT instead of VAR_INPUT and VAR_OUTPUT respectively can help reduce memory requirements.

In Section 2.3.2 the read/write restrictions on the input and output variables of POUs were detailed. This is of particular importance for FB instances:

- Input parameters (formal parameters) of an FB instance maintain their values until the next invocation. If the called FB could change its own input variables, these values would be incorrect at the next call of the FB instance, and this would not be detected by the calling POU.
- Similarly, output parameters (return values) of an FB instance maintain their values between calls. Allowing the calling POU to alter these values would result in the called FB making incorrect assumptions about the status of its own outputs.

Like normal variables, FB instances can also be made retentive by using the keyword RETAIN, i.e. they maintain their local status information and the values of their calling interface during power failure.

Finally, the relationship between FB instances and conventional data blocks (DB) will be explained.

Relationship between FB instances and data blocks.

Before calling a conventional FB, which has no local data memory (besides formal parameters), it is common practice to activate a data block containing, for example, recipe or FB-specific data. Within the FB the data block can also serve as a local data memory area. This means that programmers can use a conventional FB with individual "instance data", but have to ensure the unambiguous assignment of the data to the FB themselves. This data is also retained between FB calls, because the data block is a global "shared memory area", as shown in Example 2.15:

JU DB 14 (* global DB *)	VAR_GLOBAL FB_14 : FB_Ex; (* global instance *) END_VAR
JU FB 14 (* FB call *) ...	CAL FB_14 (* invocation of FB instance*) ...
a) Conventional DB/FB pair	b) FB instance in IEC 61131-3

Example 2.15. The use of a conventional DB/FB pair is similar to an FB instance as defined in IEC 61131-3. This topic will be discussed in more detail in Section 7.7.

This type of instantiation is restricted to function blocks and is not applicable to functions (FUNCTION).

Programs are similarly instantiated and called as instances in the Configuration as the highest level of the POU hierarchy. But this (more powerful) kind of instance differs from that for FBs, in that it leads to the creation of run-time programs by association with different tasks. This will be described in Chapter 6.

2.4.2 Re-usable and object-oriented FBs

Function blocks are subject to certain restrictions, which make them re-usable in PLC programs:

- The declaration of variables with fixed assignment to PLC hardware addresses (see also Chapter 3: "directly represented variables": %Q, %I, %M) as "local" variables is not permitted in function blocks. This ensures that FBs are independent of specific hardware. The usage of PLC addresses as global variables in VAR_EXTERNAL is, however, not affected.
- The declaration of access paths of variable type VAR_ACCESS (see also Chapter 3) or global variables with VAR_GLOBAL is also not permitted within FBs. Global data, and thus indirectly access paths, can be accessed by means of VAR_EXTERNAL.

- External data can only be passed to the FB by means of the POU interface using parameters and external variables. There is no "inheritance", as in some other programming languages.

As a result of these features, function blocks are also referred to as *encapsulated*, which indicates that they can be used universally and are free from unwelcome side effects during execution - an important property for parts of PLC programs. Local FB data and therefore the FB function do not directly rely on global variables, I/O or system-wide communication paths. FBs can manipulate such data areas only indirectly via their (well-documented) interface.

The FB instance model with the properties of "structure" and "memory" was introduced in the previous section. Together with the property of encapsulation for re-usability a very new view of function blocks appears. This can be summarised as follows:

"A function block is an independent, encapsulated data structure with a defined algorithm working on this data."

The algorithm is represented by the code part of the FB. The data structure corresponds to the FB instance and can be "called", something which is not possible with normal data structures. From each FB type any number of instances can be derived, each independent of the other. Each instance has a unique name with its own data area.

Because of this, IEC 61131-3 considers function blocks to be "object-oriented". These features should not, however, be confused with those of today's modern "object-oriented programming languages ($\rightarrow$OOP)" such as, for example, C# with its class hierarchy!

To summarise, FBs work on their **own** data area containing input, output and local variables. In previous PLC programming systems FBs usually worked on global data areas such as flags, shared memory, I/O and data blocks.

2.4.3 Types of variables in FBs

A function block can have any number of input and output parameters, or even none at all, and can use local as well as external variables.

In addition or as an alternative to making a whole FB instance retentive, local or output variables can also be declared as retentive **within** the declaration part of the FB.

Unlike the FB instance itself which can be declared retentive using RETAIN, the values of input or input/output parameters cannot be declared retentive in the FB declaration part (RETAIN) as these are passed by the calling POU and have to be declared retentive there.

For VAR_IN_OUT it should be noted that the **pointers** to variables can be declared retentive in an instance using the qualifier RETAIN. The corresponding

values themselves can, however, be lost if they are not also declared as retentive in the calling POU.

Due to the necessary hardware-independence, directly represented variables (I/Os) may not be declared as local variables in FBs, such variables may only be "imported" as global variables using VAR_EXTERNAL.

One special feature of variable declaration in IEC 61131-3 are the so-called edge-triggered parameters. The standard provides the standard FBs R_TRIG and F_TRIG for rising and falling edge detection (see also Chapter 5).

The use of edge detection as an attribute of variable types is only possible for input variables (see Section 3.5.4).

FBs are required for the implementation of some typical basic PLC functions, such as timers and counters, as these must maintain their status information (instance data). IEC 61131-3 defines several standard FBs that will be described in more detail and with examples in Chapter 5

2.5 The Function

The basic idea of a function (FUN) as defined by IEC 61131-3 is that the instructions in the body of a function that are performed on the values of the input variables result in an **unambiguous** function value (free from *side effects*). In this sense functions can be seen as manufacturer- or application-specific extensions of the PLC's set of operations.

The following simple rule is valid for functions: the same input values always result in the same output values and function (return) value. This is independent of how often or at what time the function is called. Unlike FBs, functions do not have a memory.

Functions can be used as IL operators (instructions) as well as operands in ST expressions. Like FB types, but unlike FB instances, functions are also accessible project-wide, i.e. known to all POUs of a PLC project.

For the purpose of simplifying and unifying the basic functionality of a PLC system, IEC 61131-3 predefines a set of frequently used standard functions, whose features, run-time behaviour and calling interface are standardised (see also Chapter 5).

With the help of user-defined functions this collection can be extended to include device-specific extensions or application-specific libraries.

A detailed example of a function can be found in Appendix C. Functions have several restrictions in comparison to other POU types. These restrictions are necessary to ensure that the functions are truly independent (free of any side effects) and to allow for the use of functions within expressions, e.g. in ST. This will be dealt with in detail in the following section.

2.5.1 Types of variables in functions and the function value

Functions have any number of input and output parameters and **exactly** one function (return) value.

The function value can be of any data type, including derived data types. Thus a simple Boolean value or a floating-point double word is just as valid as an array or a complex data structure consisting of several data elements (multi-element variable), as described in Chapter 3.

Each programming language of IEC 61131-3 uses the function name as a special variable within the function body in order to explicitly assign a function value.

As functions always return the same result when provided with the same input parameters, they may not store temporary results, status information or internal data between their invocations, i.e. they operate "without memory".

Functions can use local variables for intermediate results, but these will be lost when terminating the function. Local variables can therefore not be declared as retentive.

Functions may not call function blocks such as timers, counters or edge detectors. Furthermore, the use of global variables within functions is not permitted.

The standard does not stipulate how a PLC system should treat functions and the current values of their variables after a power failure. The POU that calls the function is therefore responsible for backing up variables where necessary. In any case it makes sense to use FBs instead of functions if important data is being processed.

Furthermore, in functions (as in FBs) the declaration of directly represented variables (I/O addresses) is not permitted.

How a function for root calculation is declared and called is shown in Example 2.16: return parameters include the extracted root as well as the error flag indicating an invalid root with a negative number.

```
FUNCTION  SquareRoot : INT        (* square root calculation *)
                                  (* start of declaration part *)
VAR_INPUT                         (* Input parameter *)
  VarIn    :    REAL;             (* input variable *)
END_VAR
VAR_TEMP                          (* temporary values *)
  Result   :    REAL;            (* local variable *)
END_VAR
VAR_OUTPUT                        (* output parameter *)
  Error    :    BOOL;            (* flag for root from neg. number *)
END_VAR
                                  (* start of instruction part *)
LD       VarIn                    (* load input variable *)
LT       0                        (* negative number? *)
JMPC     M_error                  (* error case *)
LD       VarIn                    (* load input variable *)
SQRT                              (* calculate square root *)
ST       Result                   (* result is ok *)
LD       FALSE                    (* logical "0" for error flag: reset *)
ST       Error                    (* reset error flag *)
JMP      M_end                    (* done, jump to FUN end *)
M_error:                          (* handling of error "negative number" *)
LD       0                        (* zero, because of invalid result in case of error *)
ST       Result                   (* reset result *)
LD       TRUE                     (* logical "1" for error flag: set *)
ST       Error                    (* set error flag *)
M_end:
LD       Result                   (* result will be in function value! *)
RET
                                  (* FUN end *)
END_FUNCTION
```

Example 2.16. Declaration and call of a function "Square root calculation with error" in IL.

2.6 The Program

Functions and function blocks constitute "subroutines", whereas POUs of type PROGRAM build the PLC's "main program". On multitasking-capable controller hardware several main programs can be executed simultaneously. Therefore PROGRAMs have special features compared to FBs. These features will be explained in this section.

In addition to the features of FBs, a PLC programmer can use the following features in a PROGRAM:

- Declaration of directly represented variables to access the physical I/O addresses of the PLC (%Q, %I, %M) is allowed,
- Usage of VAR_ACCESS or VAR_GLOBAL is possible,
- A PROGRAM is associated with a task within the configuration, in order to form a run-time program, i.e. programs are not called explicitly by other POUs.

Variables can be assigned to the PLC I/Os in a PROGRAM by using directly represented or symbolic variables as global or POU parameters.

Furthermore programs describe the mechanisms by which communication and global data exchange to other programs take place (inside and outside the configuration). The variable type VAR_ACCESS is used for this purpose.

These features can also be used at the resource and configuration levels. This is, in fact, to be recommended for complex PLC projects.

Because of the wide functionality of the POU PROGRAM it is possible, in smaller projects, to work without a configuration definition: the PROGRAM takes over the task of assigning the program to PLC hardware.

Such possibilities depend on the functionality of a programming system and will not be dealt with any further here.

A detailed example of a PROGRAM can be found in Appendix C.

The run-time properties and special treatment of a PROGRAM in the CPU are expressed by associating the PROGRAM with TASKs. The program is instantiated, allowing it to be assigned to more than one task and to be executed several times simultaneously within the PLC. This instantiation differs, however from that for FB instances.

The assignment of programs to tasks is done in the CONFIGURATION and is explained in Chapter 6.

2.7 The Execution control with EN and ENO

In the ladder diagram LD, functions have a special feature not used in the other programming languages of IEC 61131-3: here the functions possess an additional input and output. These are the Boolean input EN (Enable In) and the Boolean output ENO (Enable Out).

Example 2.17. Graphical invocation of a function with EN/ENO in LD

Example 2.17 shows the graphical representation for calling function Fun1 with EN and ENO in LD. In this example Fun1 will only be executed if input EN has the value logical "1" (TRUE), i.e. contact Lockoff is closed. After error-free execution of the POU the output ENO is similarly "1" (TRUE) and the variable NoError remains set.

With the aid of the EN/ENO pair it is possible to at least partially integrate any function, even those whose inputs or function value are not Boolean, like Fun1 in Example 2.17, into the "power flow". The meaning of EN/ENO based on this concept is summarised in Table 2.6:

EN	Explanation[a]	ENO
EN = FALSE	If EN is FALSE when calling the POU, the code-part of the function may **not** be executed. In this case output ENO will be set to FALSE upon exiting the unexecuted POU call in order to indicate that the POU has not been executed. **Note for FB**: Assignments to inputs are implementation independent in FB. The FB instance's values of the previous call are retained. This is irrelevant in FUN (no memory).	ENO = FALSE
EN = TRUE	If EN is TRUE when calling POU , the code-part of the POU can be executed normally. In this case ENO will initially be set to TRUE **before** starting the execution.	ENO = TRUE
	ENO can afterwards be set to TRUE or FALSE by instructions executed within the POU body.	ENO = individual value
	If a program or system error (as described in Appendix E) occurs while executing the function ENO will be reset to FALSE by the PLC.	ENO = FALSE (error occurred)

a   TRUE = logical "1", FALSE = logical "0"

Table 2.6. Meaning of EN and ENO within functions

As can been seen from Table 2.6, EN and ENO determine the **control flow** in a graphical network by means of conditional function execution and error handling in case of abnormal termination. EN can be connected not only to a single contact as in Example 2.17, but also with a sub-network of several contacts, thus setting a complex precondition. ENO can be similarly be evaluated by a more complex sub-network (e.g. contacts, coils and functions). These control flow operations should however be logically distinguished from other LD/FBD operations that represent the **data flow** of the PLC program.

These special inputs/outputs EN and ENO are not treated as normal function inputs and outputs by IEC 61131-3, but are reserved only for the tasks described above. Timers or counters are typical examples of function blocks in this context.

The use of these additional inputs and outputs is not included in the other IEC 61131-3 programming languages. In FBD the representation of EN/ENO is allowed as an additional feature.

The function call in Example 2.17 can be represented in IL if the programming system supports EN/ENO as implicit system variables.

If a programming system supports the usage of EN and ENO, it is difficult to convert POUs programmed with these into textual form. In order to make this possible, EN/ENO would also have to be keywords in IL or ST and would need to be automatically generated there, as they are in LD/FBD. Then a function called in LD could be written in IL and could, for example, set the ENO flag in case of an error. Otherwise only functions written in LD/FBD can be used in LD/FBD programs. The standard, however, does not make any statement about how to use EN and ENO as keywords and graphical elements in LD/FBD in order to set and reset them.

On the other hand, it is questionable whether the usage of EN and ENO is advantageous in comparison functions (std. FUN, see also Appendix A). A comparison function then has two Boolean outputs, each of which can be connected with a coil. If this comparison is used within a parallel branch of an LD network, ENO and the output Q have to be connected separately: ENO continues the parallel branch while Q, in a sense, opens a new sub-network.

Because of this complexity only some of today's IEC programming systems use EN/ENO. Instead of **dictating** the Boolean pair EN/ENO in LD/FBD there are other conceivable alternatives:

- EN and ENO can be used both implicitly and explicitly in all programming languages,
- Each function which can be called in LD/FBD must have at least one binary input and output respectively,
- Only standard functions have an EN/ENO pair (for error handling within the PLC system). This pair may not be used for user-defined functions.

The third alternative is the nearest to the definition in IEC 61131-3. This would, however, mean that EN and ENO are PLC system variables, which cannot be manipulated by the PLC programmer.

2.8 Calling Functions and Function Blocks

In this section we will deal with the special features which have to be considered when calling functions and function blocks. These features apply to standard as well as user-defined functions and function blocks.

The following examples will be given in IL. The use of ST and graphical representation in LD and FBD are topics of Chapter 4.

2.8.1 Mutual calls of POUs

The following rules, visualised in Figure 2.7, can be applied to the mutual calling of POU types:

- PROGRAM may call FUNCTION_BLOCK and FUNCTION, but not the other way round,
- FUNCTION_BLOCK may call FUNCTION_BLOCK,
- FUNCTION_BLOCK may call FUNCTION, but not the other way round,

- Calls of POUs may not be recursive, i.e. a POU may not call (an instance of) itself either directly or indirectly.

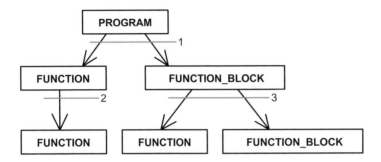

1   Program calls function or function block

2   Function calls function

3   Function block calls function or function block

Figure 2.7. The three possible ways of invocation among the POU types

Programs and FB instances may call FB instances. Functions, on the other hand, may not call FB instances, as otherwise the independence (freedom from side effects) of functions could not be guaranteed.

Programs (PROGRAM) are instantiated to form run-time programs within the Configuration by association with a TASK. They are then called by the Resource.

2.8.2 Recursive calls are invalid

IEC 1131-3 clearly defines that POUs may not call themselves ("recursion") either directly or indirectly, i.e. a POU may not call a POU instance of the same type and/or name. This would mean that a POU could "define itself" by using its own name in its declaration or calling itself within its own body. Recursion is, however, usually permitted in other programming languages in the PC world.

If recursion were allowed, it would not be possible for the programming system to calculate the maximum memory space needed by a recursive PLC program at run time.

Recursion can always be replaced by corresponding iterative constructs, i.e. by building program loops.

Both the following figures show examples of invalid calls:

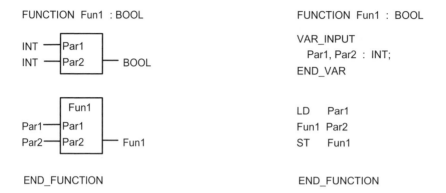

Example 2.18. Invalid recursive call of a function in graphical and IL representation: nested invocation.

In Example 2.18 the same function is called again within function Fun1.

The top half of this example shows the declaration part of the function (input variables Par1 and Par2 of data type INT and function value of type BOOL).

In the bottom part, this function is called with the same input variables so that there would be an endless (recursive) chain of calls at run time.

```
FUNCTION_BLOCK FunBlk
VAR_INPUT
  In1  :  DINT;                (* input variable *)
END_VAR
VAR
  InstFunBlk  : FunBlk;        (* improper instance of the same type *)
  Var1 :  DINT;                (* local variable *)
END_VAR
...
  CALC  InstFunBlk (In1 := Var1);   (* invalid recursive invocation! *)
...
END_FUNCTION_BLOCK
```

Example 2.19. Invalid recursive call of an FB in IL: nesting already in declaration part.

Example 2.19 shows function block FunBst, in whose local variable declaration (VAR) an instance InstFunBlk of its own type (FunBst) is declared. This instance is called in the function body. This would result in endlessly deep nesting when instantiating the FB in the declaration, and the memory space required for the instance at run time would be impossible to determine.

Programmers themselves or the programming/PLC system must check whether unintentional recursive calling exists in the PLC program.

This checking can be carried out when creating the program by means of a POU calling tree, as the invalid use of recursion applies to FB types and not to their instance names. This is even possible if FB instance names are used as input parameters (see also Section 2.8.6).

The following example shows how recursive calls can occur even if a function or FB instance does not directly call itself. It suffices if they mutually call each other.

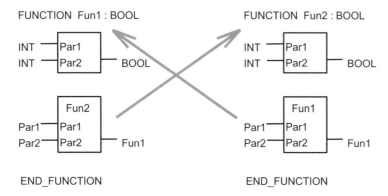

FUNCTION Fun1 : BOOL

FUNCTION Fun2 : BOOL

END_FUNCTION   END_FUNCTION

Example 2.20. Recursion by mutual invocation in graphical representation

Such types of recursion are, on principle, not permitted in IEC 61131-3. The calling condition may be defined as follows: if a POU is called by POU A, that POU and all the POUs below it in the calling hierarchy may not use the name of POU A (FB instance or function name).

Unlike most of the modern high-level languages (such as C), recursion is therefore prohibited by IEC 61131-3. This helps protect PLC programs against program errors caused by unintentional recursion.

2.8.3 Extendibility and overloading

Standard functions such as additions may have more than two input parameters. This is called input *extension* and will make use of the same function for multiple input parameters clearer. A standard function or a standard function block type is *overloaded* if these POUs can work with input data elements of various data types. These concepts are described in detail in Section 5.1.1.

2.8.4 Calling with formal parameters

When a FUN/FB is called, the input parameters are passed to the POU's input variables. These input variables are also called *formal parameters*, i.e. they are placeholders. The input parameters are known as *actual parameters* in order to express that they contain actual input values.

When calling a POU, the formal parameters may or may not be explicitly specified. This depends on the POU type (FUN or FB) and the programming language used for the POU call (see also Chapter 4).

Table 2.7 gives a summary of which POU types can be called, in textual and graphical representation, with or without giving the formal parameter names (without also called "non-formal").

Language	Function	Function block	Program
IL	with or without	with [a]	with
ST	with or without [b]	with	with
LD and FBD	with [b]	with	with

a   possible in three different ways, see Section 4.1.4
b   with std. FUN: if a parameter name exists; EN will always come first.

Table 2.7. Possible explicit specification of formal parameters ("with" or "without") in POU calls

In FBs and PROGs the formal parameters must always be specified explicitly, independently of the programming language. In IL there are different ways of doing this (see Section 4.1.4).

In ST functions can be called with or without specifying the names of the formal parameters.

Many formal parameters of standard functions do not have a name (see Appendix A.2). Therefore these cannot be displayed in graphical representation and cannot be explicitly specified in textual languages.

IEC 61131-3 does not state whether the names of formal parameters can be specified when calling user-defined functions in IL. However, in order to keep such function calls consistent with those of standard functions, it is assumed that the names of formal parameters may **not** be used with function calls in IL.

The same rules are valid for the calling of standard functions and standard function blocks. Example 2.21 shows examples for each POU type.

FB declaration:	FUN declaration:	PROG declaration:

FUNCTION_BLOCK **FBlk**
VAR_INPUT
  Par1 :   TIME;
  Par2 :   WORD;
  Par3 :   INT;
END_VAR
... (*instructions *)
END_FUNCTION_BLOCK
	FUNCTION **Fctn** : INT
	VAR_INPUT
	  Par1 : TIME;
	  Par2 : WORD;
	  Par3 : INT;
	END_VAR
	... (*instructions *)
	END_FUNCTION
		PROGRAM **Prgrm**
		VAR_GLOBAL
		  FunBlk : FBlk;
		  VarGlob : INT;
		  AT %IW4 : WORD;
		END_VAR
		... (* instructions *)
		END_PROGRAM

```
(* 1. Invocations in IL *)
LD              t#20:12
Fctn            %IW4, VarGlob                                (* function call *)
CAL FunBlk   (Par1 := t#20:12, Par2 := %IW4, Par3 := VarGlob)  (* FB call *)
```

```
(* 2. Invocations in ST *)
Fctn          (t#20:12, %IW4, VarGlob)                          (* function call *)
Fctn          (Par1 := t#20:12, Par2 := %IW4, Par3 := VarGlob); (* function call *)
FunBlk        (Par1 := t#20:12, Par2 := %IW4, Par3 := VarGlob); (* FB call *)
```

Example 2.21. Equivalent function and function block calls with and without explicit formal parameters in the textual languages IL and ST. In both cases the invocation (calling) is done in the program **Prgrm**.

In IL the first actual parameter is loaded as the current result (CR) before the invocation instruction (CAL) is given, as can be seen from the call of function Fctn in Example 2.21. When calling the function the other two parameters are specified separated by commas, the names of these formal parameters may not be included.

The two equivalent calls in ST can be written with or without the names of formal parameters. The input parameters are enclosed in brackets each time.

In the call of FB instance FunBlk in this example all three formal parameters are specified in full in both IL and ST.

The usage of formal and actual parameters in graphical representation is shown in Example 3.19.

2.8.5 Calls with input parameters omitted or in a different order

Functions and function blocks can be called even if the input parameter list is incomplete or not every parameter is assigned a value.

If input parameters are **omitted**, the names of the formal parameters that are used must be specified explicitly. This ensures that the programming system can assign the actual parameters to the correct formal parameters.

If the **order** of the parameters in a FUN/FB call is to be changed, it is also necessary to specify the formal parameter names explicitly. These situations are shown, as an example in IL, in Example 2.22.

```
(* 1.   complete FB call *)
CAL    FunBlk (Par1 := t#20:12, Par2 := %IW4, Par3 := VarGlob);

(* 2.   complete FB call with parameters in a changed order *)
CAL    FunBlk (Par2 := %IW4, Par1 := t#20:12, Par3 := VarGlob);

(* 3.   incomplete FB call *)
CAL    FunBlk (Par2 := %IW4);

(* 4.   incomplete FB call with parameters in a changed order *)
CAL    FunBlk (Par3 := VarGlob, Par1 := t#20:12);
```

Example 2.22. Examples of the FB call from Example 2.21 with parameters omitted and in a different order, written in IL

This means that either all formal parameters must be specified, and the parameter order is not relevant, or no formal parameters are used and the entries must appear in the correct order. The formal parameters always have to be specified when calling FBs, for functions this is, however, language-dependent (see Table 2.7).

Assignments to input variables can be omitted if the input variables are "initialised" in the declaration part of the POU. Instead of the actual parameter that is missing the initial value will then be used. If there is no user-defined initial value the default value for the standard data types of IEC 61131-3 will be used. This ensures that input variables **always** have values.

For FBs initialisation is performed only for the first call of an instance. After this the values from the last call still exist, because instance data (including input variables) is retained.

2.8.6 FB instances as actual FB parameters

This section describes the use of FB instance names as well as their inputs and outputs as actual parameters in the calling of other function blocks.

Using Example 2.23 this section explains what facilities IEC 61131-3 offers for indirect calling or indirect parameter assignment of FB instances.

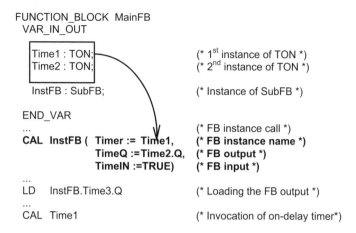

```
FUNCTION_BLOCK  MainFB
    VAR_IN_OUT

        Time1 : TON;              (* 1st instance of TON *)
        Time2 : TON;              (* 2nd instance of TON *)

        InstFB : SubFB;           (* Instance of SubFB *)

    END_VAR
    ...                           (* FB instance call *)
    CAL  InstFB ( Timer := Time1, (* FB instance name *)
                  TimeQ :=Time2.Q, (* FB output *)
                  TimeIN :=TRUE)   (* FB input *)
    ...
    LD    InstFB.Time3.Q          (* Loading the FB output *)
    ...
    CAL   Time1                   (* Invocation of on-delay timer*)
```

Example 2.23. Using the FB instance name Time1 and the output parameter Time2.Q as actual parameters of another FB. The timers Time1 and Time2 are instances of the standard FB TON (on-delay, see Chapter 5). SubFB will be declared in Example 2.24.

Instance names and the inputs and outputs of instances can be used as the actual parameters for input or input/output variables. Table 2.8 shows the cases where this is allowed (in table: "yes") and not allowed ("no").

		As actual parameter for **SubFB**		Return value, External variable
Instance	Example	VAR_INPUT	VAR_IN_OUT (pointer)	VAR_EXTERNAL VAR_OUTPUT
Instance name	Time1	yes [a]	yes [b]	yes [c]
-input	Time2.IN	-	-	-
-output	Time2.Q	yes	no [d]	-

a   Instance may not be called within SubFB (indirect FB call not possible)
b   Indirect FB call is possible, output of instance may not be changed within SubFB
c   Direct FB call, output of instance may not be changed within MainFB
d   The function (return) value of a function cannot be used either

Table 2.8. Possible cases for using or indirectly calling FB instances as actual parameters of FBs. The "Example" column refers to Example 2.23. The last column shows that FB instances may also be used as external variables or as return values. Time2.IN may not be used for read access and cannot therefore be passed as a parameter.

As this summary shows, only certain combinations of function block instance names and their inputs and outputs can be passed as actual parameters to function blocks for each of the variable types.

VAR_INPUT: FB instances and their outputs cannot be called or altered within SubFB if they are passed as VAR_INPUT. They may, however, be read.

VAR_IN_OUT: The output of an FB instance, whose pointer would be used here, is not allowed as a parameter for this variable type. An erroneous manipulation of this output can thus be prevented. Similarly, a pointer to the function value of a function is **not** allowed as a parameter for a VAR_IN_OUT variable.

The instance passed as a parameter can then be called, thereby implementing an **indirect** FB call.

The outputs of the FB instance that has been passed may not be written to. FB instance inputs may, however, be freely accessed.

VAR_EXTERNAL, VAR_OUTPUT: FB instances are called directly, their inputs and outputs may only be read by the calling POU.

Example of an indirect FB call.
Example 2.24 shows (together with Example 2.23) the use of some cases permitted in Table 2.8 within function block SubFB.

```
FUNCTION_BLOCK  SubFB
  VAR_INPUT
    TimeIN :  BOOL;          (* Boolean input variable *)
    TimeQ  :  BOOL;          (* Boolean input variable *)
  END_VAR
  VAR_IN_OUT
    Timer  :  TON;           (* pointer to instance Time1 of TON – input/output variable *)
  END_VAR
  VAR_OUTPUT
    Time3  :  TON;           (* 3rd instance of TON *)
  END_VAR
  VAR
    Start : BOOL := TRUE;    (* local Boolean variable *)
  END_VAR
  ...
```

(* Indirect call of Time1 setting/checking the actual parameter values using Timer *)		
LD	Start	
ST	**Timer.IN**	(* starting of Timer Time1 *)
CAL	**Timer**	(* calling the on-delay timer Time 1 indirectly *)
LD	**Timer.Q**	(* checking the output of Time1 *)

```
  ...
```

(* Direct call of Time3; indirect access to Time2 *)		
LD	TimeIN	(*indirect checking of the input of Time2 is not possible *)
ST	**Time3.IN**	(* starting the timer using Time3.IN *)
CAL	**Time3**	(* calling the on-delay timer Time3 directly *)
LD	**Time3.Q**	(*checking the output using Time3.Q *)
...		
LD	TimeQ	(*indirectly checking the output of Time 2 *)

```
  ...
END_FUNCTION_BLOCK
```

Example 2.24. Alternative ways of calling the on-delay FB Time1 from Example 2.23 indirectly and usage of its inputs and outputs

This example shows the *indirect call* of FB Time1, whose instance name was passed to FB SubFB as an input/output variable in Example 2.23. The function block SubFB is only assigned the FB instance name Time1 at the run time of MainFB. In SubFB Time1 (as input variable Timer) is provided with the parameter Timer.IN and then called.

As shown with Example 2.24, it is also possible to access the inputs and outputs of an FB passed as an instance name. Here the instance inputs (Timer.IN) can be read and written to, whereas the outputs (as Timer.Q) can only be read.

The FB instance Time3 in this example serves as a comparison between the treatment of input parameters and of the return values of an FB as output variables.

FB instance names as actual parameters of functions.
Instance names (such as Time1) and components of an FB instance (such as Time2.Q) can also be used as actual parameters for functions. Initially this appears to be inconsistent with the requirement that functions have to produce the same result when supplied with the same inputs and that they may not call FBs.

This is, however, not as contradictory as it seems: the FB instance passed as a parameter is not **called**, but its input and output variables are treated like elements of a normal data structure, see also Section 2.4.1.

Function values as actual parameters.
Functions and function values may also be used as actual parameters for functions and function blocks. The input variables have the same data type as the function and are assigned the function value when called.

IEC 61131-3 does not give any explicit instructions about this possibility, thus making it implementation-dependent.

Initialisation of FB instances.
Instances of function blocks store the status of input, output and internal variables. This was called "memory" above. FB instances can also be initialised, as shown in example 2.25.

```
VAR Instance_Ex :
        FunBlk (Par3 := 55, Par1 := t#20:12);
END_VAR
```

Example 2.25. Example of FB call from example 2.21 with parameters omitted and in a different order, written in IL

2.9 Summary of POU Features

The following table summarises all essential POU features that have been presented and discussed in this chapter.

Feature	Function	Function Block	Program
Input parameter	yes	yes	yes
Output parameter	yes	yes	yes
Input/output parameter	yes	yes	yes
Function value	yes	no	no
Invocation of functions	yes	yes	yes
Invocation of function blocks	no	yes	yes
Invocation of programs	no	no	no
Declaration of global variables	no	no	yes
Access to external variables	no	yes	yes
Declaration of directly represented variables [a]	no	no	yes
Declaration of local variables	yes	yes	yes
Declaration of FB instances	no	yes	yes
Overloading [b]	yes	yes	no
Extension [c]	yes	no	no
Edge detection possible	no	yes	yes
Usage of EN/ENO [c]	yes	yes	no
Retention of local and output variables	no	yes	yes
Indirect FB call	no	yes	yes
Initialisation of FB instances	no	yes	no
Usage of function values as input parameters [d]	yes	yes	yes
Usage of FB instances as input parameters	yes	yes	yes
Recursive invocation	no	no	no

a   for function blocks only with VAR_EXTERNAL
b   for standard functions
c   for standard functions and standard function blocks
d   not in IL, otherwise: implementation-dependent

Table 2.9. An overview of the POU features summarising the important topics of this chapter. The entries "yes" or "no" mean "permitted" and "not permitted" for the corresponding POU type respectively.

3 Variables, Data Types and Common Elements

This chapter presents the syntax and semantics of the basic, common language elements of all programming languages of IEC 61131-3.

The *syntax* describes the language elements IEC 61131-3 makes available for its programming languages and how they may be used and combined with each other; their meaning is governed by the *semantics*.

The first section deals with the "simple language elements" which represent the basic elements of the languages.

"Data type definition" and "variable declaration" are then explained at length.

3.1 Simple Language Elements

Every PLC program consists of a number of basic language elements or "smallest units" put together to form declarations and/or statements and finally whole programs. These *simple language elements* can be divided into:

- Delimiters,
- Keywords,
- Literals,
- Identifiers.

K.-H. John, M. Tiegelkamp, *IEC 61131-3: Programming Industrial Automation Systems*, 2nd ed., DOI 10.1007/978-3-642-12015-2_3,
© Springer-Verlag Berlin Heidelberg 2010

This division is best shown by means of a simple example:

```
FUNCTION RealAdd: REAL          (* function heading *)
  VAR_INPUT                     (* variable type "input" *)
    Inp1, Inp2: REAL;           (* variable declaration *)
  END_VAR                       (* end of variable type *)
  RealAdd := Inp1 + Inp2 + 7.456E-3;   (* ST statement *)
END_FUNCTION                    (* end of the function *)
```

Example 3.1. Function declaration with "simple language elements" in ST. Keywords are shown in bold type, identifiers in normal type and the literal in italics. Delimiters are colon, comma, parentheses, asterisk, equal, plus, minus and semicolon.

The function RealAdd describes a function for floating-point (REAL) addition of two input values Inp1, Inp2 and the constant $7.456 * 10^{-3}$. It consists of a declaration part (VAR_INPUT), in which the names and types of the two input parameters are declared, as well as a single statement line in the language Structured Text (ST).

In Example 3.1 the *keywords* are shown in bold type. This is the fixed notation specified by IEC 61131-3 for structuring declarations and statements. These keywords are the elementary "words" of the programming languages of IEC 61131-3.

The user-specific *identifiers* are shown in normal type. They are used by the PLC programmer for naming variables, functions etc.

The only *literal* employed here (in this case a "numeric literal") is printed in italics and designates a constant in floating-point representation with exponent information. The values of the data types are represented by literals, for example, numbers or character strings.

The *delimiters* in Example 3.1 are the remaining symbols and blank spaces between the other elements. These are listed in Appendix H.2 and will not be discussed here.

Table 3.1 shows further examples:

Language element	Meaning	Examples
Delimiters	Special characters with different meanings	(,),+,-,*,$,;,:=,#, space
Keywords	Standard identifiers as "words" of the programming languages	RETAIN, VAR_INPUT, END_VAR, FUNCTION
Literals	Value representations for different data types	62, 3.4509E-12, 16#a5
Identifiers	Alphanumeric character strings for user-specific variable names, labels or POUs etc.	Var_1, Inp1, EmergOff, REAL_OUT, RealAdd

Table 3.1. Examples of simple language elements

3.1.1 Reserved keywords

Keywords are standard identifiers whose spelling and intended purpose are clearly defined by IEC 61131-3.

They **cannot** therefore be employed for user-defined variables or other names.

The use of upper or lower case letters is not significant for keywords i.e. they can be represented as desired, upper case, lower case or a mixture of the two. For better distinction, the keywords in this book are generally printed in upper case.

The *reserved keywords* also include:

- Names of elementary data types
- Names of standard functions (std. FUN)
- Names of standard function blocks (std. FB)
- Names of input parameters of standard functions
- Names of input and output parameters of standard FBs
- Variables EN and ENO in graphical programming languages
- Operators in the language Instruction List
- Elements of the language Structured Text
- Language elements of the language Sequential Function Chart.

The reserved keywords defined in IEC 61131-3 are listed in Appendix H.1 and are not discussed here.

3.2 Literals and Identifiers

3.2.1 Literals

Literals represent the values of variables (constant factors). The format depends on the data types of the variables, which in turn determine the possible value ranges. Table 3.2 gives some typical examples of literals for number representation. There are three basic types of literals:

- Numeric literals (numeric values for bit string numbers as well as integers and floating-point numbers)
- Character string literals (values of character strings – in single-byte or double-byte representation)
- Time literals (values for time, duration and date).

Data type	Number representation	Remarks
		Bit string numbers
Boolean	FALSE, TRUE	Boolean representation
Byte	11, 16#0B, 2#0000_1011	Number 11 in decimal, hexadecimal and binary notation
Double word	16#ABCDEF, 16#ab_cdef	Number 11.259.375 hexadecimal
		Integers and floating-point numbers
Integer with	+3829, -45	Integer, with and without sign,
type notation	DINT#5, UINT#16#9AF,	with and without sign,
	BOOL#0, BOOL#TRUE	also Boolean
Floating point	567.82, -0.03	Real
Floating point	667E+4, -29E-16, 3e6	Real with exponent
		Character strings
Character string	"	Empty character string
Character string	'this is a text'	Non-empty character string: "this is a text"
Character string	'ÄË', '$C4$CB'	Two identical character strings in hexadecimal notation of the values of the ISO/IEC 10646-1 character set, but
Double-byte character string	"ÄË", "$00C4$00CB"	with double quotation marks
with type	STRING# 'YES',	Three single-byte characters
notation	WSTRING# "YES"	Three double-byte characters

Table 3.2. Examples of literals of different data types. As for keywords, upper case/lower case is not significant. (Continued on next page)

Data type	Number representation	Remarks
		Time, Duration and Date
Duration	t#1d2h7m19s45.7ms time#2h_7m_19s TIME#-22s150ms	Specification of days (d), hours (h), minutes (m), seconds (s) and milli-seconds (ms), also negative values
Date	d#1994-09-23	Specification of year-month-day
Time of day	tod#12:16:28.44	Specification of hours:minutes:seconds.hundredths
Date and time	dt#1994-09-23-12:16:28.44	Date and time of day combined with "-"

Table 3.2. (Continued)

Numeric and time literals may contain additional underline characters in order to give a better optical representation. Upper case/lower case is unimportant.

The most significant unit in a duration literal may "overflow", e.g. the duration value t#127m_19s is valid and the programming system makes the conversion into the "correct" representation t#2h_7m_19s.

   While a duration serves for the measurement and processing of a **relatively** elapsed time, the remaining time literals represent **absolute** times of day and dates.

Literals for times and date can be represented in short form or written out in full for ease of reading. Table 3.3 shows columns of equivalents.

Duration	Date	Time of day	Date and time
TIME#	DATE#	TIME_OF_DAY#	DATE_AND_TIME#
T#	D#	TOD#	DT#
time#	date#	time_of_day#	date_and_time#
t#	d#	tod#	dt#
Time#	dATE#	Time_of_Day#	dAtE_aNd_TiMe#

Table 3.3. Long and short form of the prefix for time and date literals.

Character string literals are represented between single quotation marks. The dollar sign ("$") is used as a prefix to enable special characters to be included in a character string. Non-printable special characters are used for formatting text for display or printout.

   Dollar signs and quotation marks themselves must therefore be identified by an additional preceeding "$".

   Table 3.4 gives a guide to the rules for special characters.

$ combination	On screen or printer
$nn	Character "nn" in hexadecimal representation
$$	Dollar sign
$', $"	Single or double quotation mark
$L, $l	Line feed (= $0A)
$N, $n	New line
$P, $p	New page
$R, $r	Carriage return (= $0D)
$T, $t	Tab
Example:	
'one $'piece$' costs $$ 45'	Character string: "one 'piece' costs $ 45"

Table 3.4. Use of the $ sign in character strings (STRING, WSTRING)

The characters immediately following a dollar sign can be written in upper or lower case.

3.2.2 Identifiers

Identifiers are alphanumeric character strings which the PLC programmer can use to assign individual names for variables, programs etc. Table 3.5 lists the language elements of IEC 61131-3 for which names can be assigned.

Jump and network labels
Enumeration constants
Configurations, Resources, Tasks/Run-time programs
Programs, Functions, Function blocks
Access paths
Variables (general, symbolic and directly represented variables)
Derived data types, Components of a structure
Transitions, Steps, Action blocks

Table 3.5. Language elements of IEC 61131-3 for which identifiers (names) can be assigned

Identifiers begin with a letter or a (single) underline character, followed by as many letters, digits and underline characters as desired.

No distinction is made between upper and lower case letters, i.e. the variable "EMERG_OFF" is identical to "Emerg_Off" or "emerg_off". The programming system assigns the same storage space to these three identifiers.

The length of identifiers is limited only by the ability of the programming system. IEC 61131-3 requires (at least) the first six characters of an identifier to be unambiguous i.e. these are **significant**.

If, for example, a programming system allows 16 characters per identifier name, the programmer must ensure that the first six characters of a name are unique to this identifier: The variable names _DRILLTOOL_8 and _DRILL are considered to be identical in a system with only six significant places. 32 or more significant places are quite usual in a modern system.

Table 3.6 gives some examples of valid and invalid identifiers.

Valid identifiers	Invalid identifiers
MW2	2MW
VALVE3X7, Valve3x7	3X7
EMERG_OFF, Emerg_Off	Emerg Off
EMERGOFF, EmergOff	
DRILLTOOL_8, drilltool_8	_DRILL
DRILLTOOL, _DRILL	
3ST91	3ST9$1, 3ST9.1

Table 3.6. Examples of valid and invalid (crossed out) identifiers

3.2.3 Comments

Comments may be used wherever empty spaces are allowed, except in string character literals. They are introduced by the two characters "left parenthesis" and asterisk "(*" and concluded symmetrically with "*)".

Comments must not be nested and have no syntactic or semantic significance to declarations or one of the languages defined by IEC 61131-3.

3.2.4 Pragmas

The standard explicitly allows so-called *pragmas*, which are typically used for automatic pre-processing and post-processing of programs. These elements are identified by braces. Their syntax and semantics depend on the implementation by the programming system. They are therefore not defined by the standard itself.

Pragmas may be used wherever comments are allowed.

{Author MT, Version D4.1}
{n := 4}

Example 3.2. Example of pragmas that are not defined by the standard itself.

3.3 Meanings of Data Types and Variables

Variables are declared in the declaration part of a POU, i.e. they are "introduced" together with their properties (see also Sections 2.3.2 and Example 2.10). Variable declarations are independent of the chosen programming language and therefore uniform for the entire PLC project. Declarations essentially consist of an identifier (variable name) as well as information about the data type used. Type definitions contain the application-specific (derived) data types and are valid project-wide.

Before the practical use of data types and variables can be explained in depth (in Sections 3.4 and 3.5), the meaning of these PLC programming terms, which are new for the classical PLC world, will be explained.

3.3.1 From direct PLC addresses via symbols to variables

In conventional PLC programming (see DIN 19239) it is normal to access addresses in PLC memory directly using "operands", such as "M 3.1" (flag, or memory bit 3.1) or "IW 4" (input word 4). These addresses can either be in the main memory of the PLC central processing unit (CPU) or, for example, in the *I/O modules* (inputs and outputs). Addresses are typically accessed as bit, byte, word or double word.

The memory areas addressed by physical addresses can be used for different purposes in the PLC program: as integer or BCD value (e.g. BYTE or WORD), as a floating-point number (e.g. REAL, LREAL) or as a timer or counter value (e.g. INT), etc. This means that the memory cell has a specific data format in each case (8, 16, 32 bits). These data formats are in general incompatible with each another and programmers have to remember in which format the PLC addresses in a program may be used.

Erroneous programs may result when an incorrect memory address is specified (address areas of 16KB and more are often available) or an address is used in the wrong data format.

For many PLC systems "symbols", which can be used equivalently in place of the absolute PLC addresses, were therefore introduced to ensure a more readable PLC program. Every address is assigned a unique symbolic name by means of an assignment list or symbol table. This use of symbolic representation as a "variable substitute" is shown on the left-hand side of Example 3.3.

IEC 61131-3 goes one step further: in place of the hardware addresses or symbols, the use of *variables* is defined, as is normal in high-level programming languages. Variables are identifiers (names) assigned by the programmer, which act as "place-holders" and contain the data values of the program.

Example 3.3 shows the PLC addresses and symbols on the left-hand side compared with the corresponding IEC 61131-3 declarations on the right-hand side.

PB 123:   PROGRAM **ProgIEC**

Assignment list (symbol table):	VAR
I3.4 = **InpVar**	**InpVar** AT %IX3.4 : BOOL;
M70.7 = **FlagVar**	**FlagVar** : BOOL;
Q1.0 = **OutVar**	**OutVar** AT %QX1.0 : BOOL;
	AT %MX70.6 : BOOL;
	END_VAR

A InpVar	LD InpVar
A FlagVar	AND FlagVar
ON M70.6	ORN %MX70.6
= OutVar	ST OutVar
...	...

Example 3.3. Introduction of the terms "variable" and "data type" by IEC 61131-3. On the left-hand side of the bottom box is a simple IL (=STL) program section according to DIN 19239, on the right-hand side is the corresponding code using IEC 61131-3.

As already described in Chapter 2, every variable of a POU is declared in a declaration block, i.e. it is defined with all its properties. In Example 3.3 (on the right-hand side) the variable type VAR is used for declaration of local variables. The use of the keyword AT shown here, as well as addresses beginning with "%" are explained in Section 3.5.1.

The symbols or variables InpVar and OutVar in Example 3.3 are addressed directly as hardware addresses "I 3.4" and "Q 1.0" of the PLC. The PLC address "M 70.6" is used directly without a symbol or variable name. All the addresses and variables in this example are "Boolean" (with binary values 0 and 1). Therefore the data type "BOOL" (abbreviation for Boolean) is specified on the right in each case.

The variable FlagVar is declared on the right **without** direct assignment to a PLC address. The programming system does this automatically when compiling the program, by finding and assigning a free memory address (such as M 70.7).

3.3.2 The data type determines the properties of variables

IEC 61131-3 uses variables not only for PLC addresses but uniformly for all user data of the PLC program, particularly for data that need *not* be at a specific memory location or PLC address ("general" variables).

Variables have properties which are determined by the so-called *data type* assigned to them. While the variable name corresponds to the storage space of a variable, the data type indicates **which values** the variable can have.

Data types determine variable properties such as initial value, range of values or number of bits (data width). A variable is declared by assignment of a data type to an identifier (variable name) and is thereby made known to the POU and to other POUs.

In Example 3.4 the variable Start is declared as a local variable with the properties of the data type BYTE. BYTE is a standard data type of IEC 61131-3 ("elementary data type"), see also Appendix D. A variable like Start declared with BYTE has the initial value 0, a value range of 0 to 255 and occupies 8 bits.

```
VAR
  Start  :  BYTE;        (*declaration of variable "Start" with data type BYTE *)
END_VAR
```

Example 3.4. A simple variable declaration, consisting of the identifier Start (variable name) and the colon, followed by the elementary data type BYTE. It is concluded with a semicolon.

The properties of variables also depend on additional information in their declaration and on properties of the variable type in whose block they are declared.

3.3.3 Type-specific use of variables

When a variable is accessed, the programming system can check the type-specific use of the variable i.e. whether the variable is processed in accordance with its data type. This is a significant advantage compared to previous PLC programming where such checks were system-specific and could only be partially carried out, if at all.

This *type checking* is carried out automatically by the programming system while compiling the PLC program. The programmer can be warned, for example, if a variable of type BYTE (such as Start in Example 3.4) is assigned a value of the type REAL (floating-point number).

Because the properties of a variable are determined by its data type, errors caused by incorrect use of the data format can be avoided to a large extent (see also Section 3.3.1).

A further example can be seen in Example 3.5 where counter values are declared as typical variables of type integer with or without signs.

```
CounterBackward    :    INT;      (* signed integer *)
CounterForward     :    UINT;     (* unsigned integer *)
```

Example 3.5. Use of the predefined data types "(un)signed integer" for the declaration of counter variables

In Example 3.5 the two variables CounterBackward and CounterForward are assigned the data types INT and UINT. This establishes that CounterBackward can have values from -32768 to 32767 and CounterForward has the range of values between 0 and 65535.

A programming system at least issues a warning if, for example, these two variables are used together in a logic operation, e.g. comparison or addition.

3.3.4 Automatic mapping of variables onto the PLC

For many variables of a POU (temporary data etc.) it is unimportant in which PLC memory area they are kept, as long as enough storage space is available. As already described in Section 3.3.1, an explicit "manual" memory division is necessary in the case of many previous programming systems. This can lead to errors, especially with complex computations and/or large memory areas.

Using the variable concept of IEC 61131-3, such "general" variables, which were previously managed manually in the global (non-local) "flag area" of the PLC, are automatically mapped onto a corresponding storage space in the PLC during compilation of the program. The programmer does **not** have to take care of assigning the variables to physical addresses. This corresponds to the procedure used by compilers for normal high-level programming languages.

The use of general variables such as FlagVar in Example 3.3 or the variables in Examples 3.4 and 3.5 is simpler and more reliable than the direct use of memory addresses in the PLC. An inadvertent double or incorrect assignment of memory areas by the programmer is automatically excluded.

3.4 Data Types

Traditional PLC programming languages contain data types such as floating-point representation, BCD code or timer and counter values, which often have completely incompatible formats and coding.

For example, floating-point representation, for which the keywords REAL, FLOAT etc. are used, is typically implemented by 32-bit data words, and different value ranges are additionally employed for fraction and exponent.

Most traditional programming systems have a uniform use of BIT, BYTE, WORD and DWORD. However, even for simple integer values, there are fine but distinctive differences (with/without sign, number of bits) between the PLC systems of different manufacturers.

Therefore, in most cases program porting with incompatible data types requires large programming modifications, which are highly error-prone.

As a result of IEC 61131-3, the most common data types used in PLC programming are defined so that their meaning and use within the PLC world are uniform. This is of particular interest to machine and plant builders and engineering offices who work with several PLC and programming systems from different manufacturers. Uniform data types are the first step towards portable PLC programs.

3.4.1 Elementary data types

In IEC 61131-3 there is a set of predefined, standardised data types called *Elementary data types*, which are summarised in Table 3.7, see also Appendix D.

Boolean/ stringbit	Signed integer	Unsigned integer	Floating-point (Real)	Time, duration, date and character string
BOOL	INT	UINT	REAL	TIME
BYTE	SINT	USINT	LREAL	DATE
WORD	DINT	UDINT		TIME_OF_DAY
DWORD	LINT	ULINT		DATE_AND_TIME
LWORD				STRING

Meaning of the first letters: D = double, L = long, S = short, U = unsigned

Table 3.7. The elementary data types of IEC 61131-3. Their names are reserved keywords.

The elementary data types are characterised by their data width (number of bits) as well as their possible value range. Both values are defined by the IEC.

Exceptions to this rule are the data width and range of date, time and string data types which are implementation-dependent.

In the standard, neither BCD data types nor counter data types are defined. BCD code is not nearly as important now as in the past and must therefore be defined individually for special purposes in a PLC system. Counter values are implemented by normal integers, no specific format is required at least for the standard counter function blocks of IEC 61131-3.

In Appendix D all data types are listed together with their properties (range, initial values). Examples of data types have already been given in Table 3.2.

3.4.2 Derived data types (type definition)

On the basis of the elementary data types, PLC programmers can create their own, "user-defined" data types. This procedure is known as *derivation* or *type definition*. This enables programmers to implement the data model most favourable for their application.

Such type definitions are global for a PLC project. The data types defined with new names are called *derived data types* and are used for variable declarations in the same way as the elementary data types.

Textual representation has to be employed for type definitions. IEC 61131-3 does not mention graphical representation.

Type definitions are framed by the keywords TYPE ... END_TYPE, as shown in Example 3.6.

```
TYPE
   LongFloatNum  : LREAL;          (* direct derivation from IEC data type *)
   FloatingPoint : LongFloatNum;   (* direct derivation from a user-defined data type *)
   InitFloatNum  : LREAL := 1.0;   (* derivation with new initial value *)
   tControl      : BOOL := TRUE;   (* derivation with new initial value *)
END_TYPE
```

Example 3.6. Example of simple type definitions: "direct derivation" of a data type

In Example 3.6 the new data type LongFloatNum is defined as an alternative for the standard data type LREAL. After this declaration, LongFloatNum and LREAL can be used equivalently for further variable declarations.

As the example shows, a derived data type can in turn serve as the basis for a further derivation. Therefore FloatingPoint is also equivalent to LREAL. The derived data type InitFloatNum has a different initial value to LREAL, 1.0 instead of 0.0. Furthermore tControl has the initial value TRUE as opposed to the standard initial value FALSE.

Type definitions are required in order to create **new** data types with extended or different properties which can be passed on by **repeated** use.

A task to be programmed can be implemented more effectively by the use of such application-orientated data types. Customers and PLC manufacturers can create or predefine individual data types:

- Initial values deviating from the standard,
- Data types for range and enumeration,
- Multidimensional arrays,
- Complex data structures.

These possibilities can be combined with each other and are supported by IEC 61131-3. They are explained further in the following sections.

Additional properties for elementary data types.
The following additional properties can be assigned to an elementary data type as shown in Table 3.8.

Property	Meaning
Initial value	The variable is given a particular initial value.
Enumeration	The variable can assume one of a specified list of names as a value.
Range	The variable can assume values within the specified range.
Array	Several elements of the same data type are combined into an array. While accessing the array the maximal permissible subscript (index) must not be exceeded.
Structure	Several data types are combined to form one data type. A structured variable is accessed using a period and the component name.

Table 3.8. Additional properties for elementary data types

The properties "array" and "structure" can also be applied to derived data types, i.e. they can be nested. Multiple arrays of a particular data type form a multi-dimensional array type.

The "range" property is defined in IEC 61131-3 only for the elementary data type Integer and its direct derivatives. An extension to further data types is conceivable.

An enumerated data type is not a derivative in the true sense since it is not derived from any elementary data type. But in IEC 61131-3 it is defined as such, as the programming system typically uses integers to implement enumerated data types, thus giving the impression of derivation.

```
TYPE
    Colour    :   (red, yellow, green);          (* enumeration *)
    Sensor    :   INT (-56..128);                (* range *)
    Measure   :   ARRAY [1..45] OF Sensor;       (* array *)
    TestBench :                                   (* structure *)
        STRUCT
            Place  :   UINT;                      (* elementary data type *)
            Light  :   Colour:= red;             (* enumerated data type with initial value *)
            Meas1  :   Measure;                   (* array type *)
            Meas2  :   Measure;                   (* array type *)
            Meas3  :   Measure;                   (* array type *)
        END_STRUCT;
END_TYPE
```

Example 3.7. Examples of elementary data types with additional properties as derived data types

The examples in Example 3.7 show the use of these additional properties: Colour can be the range of values for a traffic light with three colours, Sensor contains a permissible temperature range and the array Measure is suitable for 45 single measurements. TestBench is a data structure made up of elementary and derived data types.

Compliance with the properties in Table 3.8 for range and array subscript (index) can be checked both statically (by the programming system) and dynamically (at run time).

The assignment of properties helps in the detection of bugs while creating programs and at run time, leading to more secure programs as well as improving the program documentation.

The three elements in parentheses for the *enumerated data type* are entered as names by the programmer without any further information being necessary. They can therefore be understood as "text constants".

The programming system automatically converts the three values red, yellow and green, for example, into suitable code. The values are usually internally mapped, invisibly for the programmer, to integer values, for example, beginning with "1". In the program the names of the colour values can be employed directly as constants. In Example 3.7 the variable for Light receives "red" as the initial value.

The use of enumerated data types simplifies automatic checking by the programming system. It also makes the programs easier to read.

If a *range* is declared for a data type, as for Sensor in Example 3.7, an error is reported if this range is exceeded during programming or at run time.

Ranges can also be used in CASE statements in the ST language in order to carry out range-dependent tasks, see Section 4.2.6.

Arrays.

Arrays are directly consecutive data elements of the same data type in memory. An array element can be accessed with the aid of an array subscript (index) within the specified *array limits*. The value of the subscript indicates which array element is to be addressed. This is illustrated by Figure 3.1.

Powerful PLC systems ensure that an error message will be issued at run time if an attempt is made to access an array with an array subscript outside the allowed array limits.

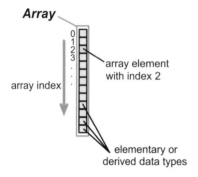

Figure 3.1. Illustration of the elements of a one-dimensional array

The array shown in Figure 3.1 is one-dimensional i.e. it has precisely one set of array limits. Multidimensional arrays may also be created by declaring further sets of array limits separated by commas, as shown in Example 3.8 (Meas_2Dim). In this case the elements are stored in memory one dimension after the other. The dimensions are specified in order of significance.

```
TYPE
  Meas_1Dim  : ARRAY [1..45]     OF Sensor;   (* 1-dimensional array *)
  Meas_2Dim  : ARRAY [1..10,1..45]OF Sensor;  (* 2-dimensional array *)
END_TYPE
```

Example 3.8. Type definitions of a one-dimensional and two-dimensional array for the acquisition of one (Meas_1Dim) or of ten (Meas_2Dim) logged measurements.

Arrays of FB instance names are not allowed in IEC 61131-3. This would nevertheless be a reasonable extension of the standard in order to enable, for example, easier access to similar timers and counters.

In addition to the definition of arrays as a data type, arrays can also be defined directly with the variable declaration. This is explained in Section 3.5. Examples of the practical use of variables of these data types are also given.

Data structures.
With the aid of the keywords STRUCT and END_STRUCT, *data structures* - known and used in high-level programming languages (e.g. struct { } in the programming language C) - can be built up hierarchically. These may contain any elementary or derived data types as sub-elements. FB instance names are similarly not allowed in data structures, although there is room here for a possible extension to IEC 61131-3.

If a sub-element is also a structure, a structure hierarchy is created (as illustrated by Figure 3.2), whose lowest structure level may consist of elementary or derived data types.

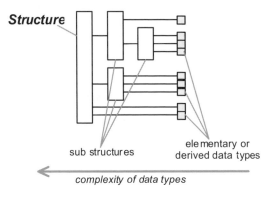

Figure 3.2. Illustration of a structure (STRUCT). It consists of (multiple) derived and/or elementary data types. The complexity of the data types increases from right to left.

In this way PLC programmers can optimally adapt their data structures to meet their requirements.

In Example 3.9 the structure MotorState created to reflect the real operating condition of a controlled motor is defined. For the definition elementary data types (including Revolutions with a range specification) as well as an enumerated type (Level) are used.

```
TYPE
   LimitedRevol   :   UINT (0..230);
   TypLevel       :   (Idling, SpeedUp1, SpeedUp2, MaxPower);
   MotorState :
   STRUCT
      Revolutions  :   LimitedRevol;        (* range *)
      Level        :   TypLevel;            (* enumerated data type *)
      MaxReached :     BOOL;                (* elementary data type *)
      Failure      :   BOOL;                (* elementary data type *)
      Brake        :   BYTE;                (* elementary data type *)
   END_STRUCT;
END_TYPE
```

Example 3.9. Type definition of a complex structure as a derived data type based on elementary data types as well as on range specification and enumeration

The new data type MotorState can be used for declaring corresponding variables. For example, in order to describe several motors of the same type an array can be formed (as a variable or also as a data type) consisting of elements with type MotorState.

```
TYPE
   MultiMotState      :  ARRAY [1..4] OF MotorState;   (* further derivation *)
END_TYPE

VAR
(* case 1: *)
   Motor1       :   MotorState;                  (* declaration *)
(* case 2: *)
   Motors       :   ARRAY [1..4] OF MotorState;  (* array declaration *)
(* case 3: *)
   FourMotors   :   MultiMotState;               (* declaration *)
END_VAR
```

Example 3.10. Use of a derived data type for further derivation (top) and array declaration (bottom)

As Example 3.10 shows, the array with the four elements of type MotorState can be created by a variable declaration: Motors has four elements (case 2).

Case 3 represents an alternative: here the data type MultiMotState is used. This data type reserves for each variable (such as FourMotors) an array of the same size consisting of four elements based on the further derived data type MotorState.

The derivation of data types can in principle be nested, i.e. data types can be derived from derived data types. However nesting is illegal if it causes recursion. This can occur, for example, when within a structure another structure already containing this structure is defined.

Example 3.11 shows a case of illegal nesting.

```
TYPE
  StructureA :
  STRUCT
    Element1 : INT;
    Element2 : StructureB;(* legal: sub structure *)
    Element3 : BYTE;
  END_STRUCT;
  StructureB :
  STRUCT
    Element1 : LINT;
    Element2 : StructureA;(* illegal: recursive to StructureA *)
    Element3 : WORD;
  END_STRUCT;
END_TYPE
```

Example 3.11. Illegal nesting of derived data types: StructureA has a substructure that contains itself (recursive type definition).

Section 3.5 gives further examples of the practical use of relevant variables.

Initial values in type definitions.
When defining a data type the programmer can assign initial values which can automatically be inserted in the relevant variable declaration. If, as in Example 3.6, values other than the default initial values specified in IEC 61131-3 are explicitly used in a variable declaration (even for individual structure elements), these values will be used in preference by the programming system.

The initial values for the data types of IEC 61131-3 are given in Appendix D.

```
TYPE
  MotorState :
  STRUCT
    Revolutions  :   LimitedRevol := 0;
    Level        :   TypLevel := Idling;
    MaxReached :     BOOL := FALSE;
    Failure      :   BOOL := FALSE;
    Brake        :   BYTE := 16#FF;
  END_STRUCT;
END_TYPE
```

Example 3.12. Type definition of the structure in Example 3.9 with assignment of initial values (shown in bold type)

In Example 3.12 the structure elements of the derived data type MotorState from Example 3.9 are provided with initial values.

For initial values of arrays, a series of value assignments can be abbreviated by specifying a repeat factor and a sub-series in brackets.

In the case of data type STRING an initial character string may be specified in single quotation marks.

```
VAR
  Field1 : ARRAY [1..12] OF SINT := [3,16#FF,-5,-9,-5,-9,-5,-9,-5,-9,0,0];
  Field2 : ARRAY [1..12] OF SINT := [3,16#FF, 4(-5,-9), 2(0)];
  Text : STRING [4] := 'stop';
END_VAR
```

Example 3.13. Example of initialisation of array variables and character strings

In Example 3.13 the two arrays Field1 and Field2 are initialised with the same initial values. For Field2 the shortened form with a repeater before the sub-series (-5,-9) and (0) is used.

3.4.3 Generic data types

IEC 61131-3 defines so-called *generic data types* in order to hierarchically combine the elementary data types into individual groups. These data types begin with the prefix ANY; e.g. all integer data types (INT) are designated as ANY_INT.

Many standard functions of IEC 61131-3 can be applied to more than one data type. Addition (ADD) can be carried out, for example, with all types of integers (INT) i.e. ADD supports the generic data type ANY_INT.

Generic data types are used to describe standard functions in order to specify which of their input or output variables allow several data types. This concept is called *overloading of functions* and is described in Section 5.1.1.

Table 3.9 shows how elementary data types are assigned to generic data types. The data type ANY here forms the widest generalisation of a data type.

If, for example, the standard function for multiplication (MUL) supports the generic data type ANY_NUM, then the programming system permits all data types for MUL which are included in the data types ANY_INT and ANY_REAL i.e. all integers with and without signs as well as floating-point numbers.

ANY_DERIVED			
ANY_ELEMENTARY			
ANY_BIT	**ANY_MAGNITUDE**	**ANY_DATE**	**ANY_STRING**
BOOL BYTE WORD DWORD LWORD		DATE TIME_OF_DAY DATE_AND_TIME	STRING WSTRING

ANY_MAGNITUDE			
ANY_NUM			
ANY_INT		**ANY_REAL**	
INT	UINT	REAL	TIME
SINT	USINT	LREAL	
DINT	UDINT		
LINT	ULINT		

Table 3.9. Overview of the generic data types ANY

The user-defined (derived) data types are also covered by the type *ANY*. For directly derived data types the generic data type of a variable is the same as that of the derived data type. In Example 3.6 the two data types LongFloatNum and FloatingPoint have the same data type LREAL.

Generic data types are used to explain the calling interface of (standard) functions and belong to the reserved keywords. Their use in user-defined POUs for variable declaration, for example, is illegal or at least not covered by the standard.

3.5 Variables

As already described in Section 3.3, variables are declared together with a data type as placeholders for application-specific data areas. Their properties can be defined in the declaration by means of:

- Properties of the specified (elementary or derived) data type,
- Information about additional initial values,
- Information about additional array limits (array definition),
- Variable type of the declaration block in which the variable is declared (with attribute/qualifier).

An example of the most important elements of a *variable declaration* is given in Example 3.14.

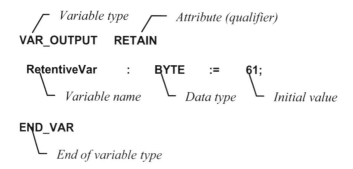

Example 3.14. Elements of a variable declaration with initial value assignment

In this example, the variable RetentiveVar of the data type BYTE is declared with the initial value 61. Because it has the qualifier RETAIN it is stored in the battery-backed part of the PLC memory.

The declaration of an *instance name* for *function blocks* represents a special case of variable declaration: An FB instance name is declared just like a variable, except that the FB type name is specified in place of the data type, as already described in Section 2.4.1.

As in the case of type definitions, initial values and arrays can also be defined at the time of declaration with an "unnamed" type definition (see examples in Section 3.5.2). IEC 61131-3 does not provide this facility for the properties enumeration, range and structure.

3.5.1 Inputs, outputs and flags as special variables

The familiar PLC terms inputs, outputs and flags are given special treatment in the IEC variable concept. A short introduction has already been given in Section 3.3.1.

In order to **directly** access the data areas of the PLC system's processors and their I/O modules in the program, IEC 61131-3 offers the PLC programmer two possibilities:

- Directly represented variables,
- Symbolic variables.

In the declaration of such variables the physical memory location (PLC address, e.g. the *I/O module address*) is specified with the keyword **AT**. The address structure is shown in Table 3.10. These direct PLC addresses are also called *hierarchical addresses.*

They begin with a "%", followed by a letter I (input), Q (output) or M (flag/memory). This is followed by another letter, which indicates the data width of the PLC address. "X" (bit address) can also be omitted.

Direct PLC addresses				Explanations
%				introductory character
	I			input
	Q			output
	M			flag/memory
		none		
		X		bit
		B		bit (optional)
		W		byte
		D		word
		L		double word
		*		long word
				memory location, not (yet) defined
		v.w.x.y.z		multi-digit hierarchical address, increasing in significance from right to left. The number and interpretation of the places are dependent on the manufacturer, e.g.: z - bit, y - word, x - module, w - bus, v - PLC
				Examples
%	I	W	7	input word 7
%	Q	D	3.1	output double word 1 in module 3
%	M		5.2.0	flag 0 of word 2 of module 5
%	M	X	5.2.0	flag 0 of word 2 of module 5
%	I		80	input bit 80
%	Q	B	4	output byte 4
%	Q	*		output at storage location that has not yet been defined

Table 3.10. Structure of the direct PLC addresses with examples. They are used for the declaration of "directly represented" and "symbolic" variables.

In the case of *directly represented variables*, a data type is assigned to a hierarchical address (see Example 3.15). The declaration of *symbolic variables* also contains a variable name with which the desired input, output or flag can be accessed "symbolically".

```
VAR
  (* directly represented variable *)
  AT %IW6  : WORD;                         (* input word starting at address 6 *)
  AT %QD3  : DINT;                         (* output double word starting at address 3 *)
  (* Symbolic variable *)
  OUT_HG   AT %QW7:      WORD;             (* output word 7 at address 7*)
  AD_3      AT %QD3:      DINT;            (* output double word at address 3 *)
  C2 AT %Q*:             BYTE;            (* assignment C2 to output storage location*)
  END_VAR

  ...
  LD        %IW6                          (* use of a directly represented variable *)
  ST        OUT_HG                        (* use of a symbolic variable *)
  ...
```

Example 3.15. Examples of directly represented and symbolic variables and their use in IL with the aid of hierarchical addresses

Assigning a data type to a flag or I/O address enables the programming system to check whether the variable is being accessed correctly. For example, a variable declared by "AT %QD3 : DINT;" cannot be inadvertently accessed with UINT or REAL.

Directly represented variables replace the direct *PLC addresses* often employed in programs up until now (e.g. I 1.2). In this case, the address also acts as the variable name (like %IW6 in Example 3.15).

Symbolic variables are declared and used in the same way as normal variables, except that their storage location **cannot** be freely assigned by the programming system, but is restricted to the address specified by the user with "AT" (like OUT_HG in Example 3.15). These variables correspond to addresses which were previously assigned symbolic names via an *assignment list* or *symbol table*.

When an asterisk is used in programs and functions blocks to mark undefined storage locations, this must be done in configurations using VAR_CONFIG ... END_VAR, see also Chapter 6.

Directly represented and symbolic variables may be declared for the variable types VAR, VAR_GLOBAL, VAR_EXTERNAL and VAR_ACCESS in programs, resources and configurations. In function blocks they can only be imported with VAR_EXTERNAL.

3.5.2 Multi-element variables: arrays and structures

IEC 61131-3 designates arrays and structures as *multi-element variables*. Accordingly, simple variables are designated as *single-element variables*. The type definitions of the multi-element variables have already been discussed in Section 3.4.2. Their use is described below.

Example 3.16 shows arrays and structures employing the type definitions of Example 3.9 and Example 3.10.

```
TYPE
  LineState  :
  STRUCT
     Running    :    BOOL;
     Drive      :    MultiMotState;
  END_STRUCT;
END_TYPE
VAR
  Input AT %IB0 : ARRAY [0..4] OF BYTE;
  Index        :    UINT := 5;
  Motor1       :    MotorState;
  FourMotors   :    MultiMotState;
  MotorArray   :    ARRAY [0..3, 0..9] OF MotorState;
  Line         :    ARRAY [0..2] OF LineState;
END_VAR
```

Example 3.16. Examples of multi-element variable declarations

Data elements of an array are accessed by selecting the integer array subscript (e.g. Index) in square brackets. Structure elements are addressed by specifying the structure's variable name, followed by a period and the name of the *structure component*.

Table 3.11 shows access examples to single-element and multi-element variables from Example 3.16.

Access to array elements	Remarks
Input [0]	first input element
FourMotors [Index]	the 4th structure, if index is 4
MotorArray [Index, 2]	34th structure MotorState, if index is 3
Access to structure elements	
Motor1.Revolutions	component Revolutions
FourMotors[1].Revolutions	component Revolutions of 2nd structure
MotorArray[Index, 2].Level	
Line[1].Drive[1].Revolutions	

Table 3.11. Examples of use of the multi-element variables in Examples 3.16, 3.9 and 3.10 - access to arrays and structures

In Example 3.10 an array was defined within a data type. As Example 3.16 shows in the case of Input or the two-dimensional array MotorArray, an array definition can also be specified within a variable declaration.

Array definition is not defined or planned within IEC 61131-3 for the enumerated data type or for ranges or *data structures*, but this would be a helpful extension to the standard.

As Table 3.11 shows, the accesses with array subscript and period can be nested for complex variables with substructures, as illustrated by the variable Line.

Initial values can be defined in variable declarations for single- and multi-element variables according to the same rules as shown in Example 3.12 and Example 3.13 for type definitions.

3.5.3 Assignment of initial values at the start of a program

Variables are assigned initial values when the resource or the configuration is started (see also Chapter 6). These initial values are dependent on the information specified by the PLC programmer in the corresponding declaration part for the variable or on the values specified for the relevant data type.

Since elementary data types (and therefore also those derived from them) have predefined default initial values, it is guaranteed that **every** variable has defined initial values.

A variable can assume initial values according to the rules shown in Table 3.12. However, the extent to which these rules can be used is typically dependent on the implementation by the programming system.

Default	Prior.[a]	Program start	Remarks
Battery-backed with RETAIN	1	**Warm restart:** Restoration of the values after power recovery or after stop (*warm reboot*).	
Initial value in declaration	2	**Cold restart:** Initial values for a defined new start of the	These values are specified in the variable declaration.
Initial value from data type	3	program, as well as after program load.	Elementary data types are initialised with 0 for numerical values and times, 01-01-01 for date and an empty character string for strings. For derived data types individual initial values can be used in the type definition.

a   Priority: 1=highest priority, 3= lowest priority

Table 3.12. Assignment of initial values according to priorities on warm and cold restart. The three possible ways of assigning initial values to variables are shown in the left-hand column.

If several methods of assigning initial values to variables are used, the one with the highest priority applies at a warm or cold restart. On warm restart a retained value has priority over the initial values specified in the variable declaration or type definition. The initial value for the data type is only employed if neither retentiveness nor an initial value is specified in the variable declaration.

If a PLC program is started by a *warm restart* (power recovery), its retentive variables retain the values they held before the interruption occurred. This behaviour is also known as *warm reboot.*

After a *cold restart* (after loading of the program into the PLC or after a stop through errors or the user), the variables are set to the initial values assigned in the definition of the data type (predefined or user-defined) or in the (user-defined) variable declaration (*New start*).

The initialisation of the inputs and outputs of a PLC system (I/O modules) and other memory areas is implementation-dependent.

Initial values are permissible for every variable type except for VAR_IN_OUT and VAR_EXTERNAL. External variables are initialised where they are declared as global - in the relevant VAR_GLOBAL. In the case of VAR_IN_OUT, initialisation is not allowed since this variable type declares pointers to variables and not the variables themselves.

3.5.4 Attributes of variable types

IEC 61131-3 defines *attributes*, or *qualifiers*, with which additional properties can be assigned to variables:

- RETAIN   Retentive variable (battery back-up)
- NON_RETAIN   Non-retentive variable (no battery back-up)
- CONSTANT   Constant variable (cannot be modified)
- R_EDGE   Rising edge
- F_EDGE   Falling edge
- READ_ONLY   Write-protected
- READ_WRITE   Can be read and written to

The RETAIN, NON_RETAIN and CONSTANT qualifiers in IEC 61131-3 are specified immediately after the keyword of the variable type. This means that these three qualifiers always refer to the entire section of the variable declaration (up to END_VAR).

The four other attributes, or qualifiers, are assigned individually for individual *variable declarations* and **cannot** be combined with the other three qualifiers.

Table 3.13 shows the variable types for which these attributes/qualifiers are permissible.

Variable type	RETAIN NON_RETAIN	CONSTANT	R_EDGE, F_EDGE	READ_ONLY, READ_WRITE
VAR	yes	yes	no	no
VAR_INPUT	yes	no	ja	no
VAR_OUTPUT	yes	no	no	no
VAR_IN_OUT	no	no	no	no
VAR_EXTERNAL	no	yes	no	no
VAR_GLOBAL	yes	yes	no	no
VAR_ACCESS	no	no	no	yes
VAR_TEMP	no	yes	yes	no
VAR_CONFIG	no	no	no	no

Table 3.13. Use of attributes/qualifiers for variable types

RETAIN is used to indicate retentive variables, i.e. variables whose values are to be retained during a loss of power. **NON_RETAIN** has the opposite effect, i.e. variables are explicitly **not** to be retained. The state of variables without the attribute **RETAIN** or **NON_RETAIN** after a loss of power is implementation-dependent, see also Table 3.12.

The attributes **RETAIN** and/or **NON_RETAIN** are permitted for VAR, VAR_INPUT, VAR_OUTPUT and VAR_GLOBAL and meaningful for retaining variables in an instance of function blocks or programs or retaining instances of structured variable types. They are, however, not permitted for **individual** elements of structures.

CONSTANT describes "variables" whose values are not allowed to be changed during program execution, i.e. they are to be treated as write-protected constants (not variables).

The qualifier **CONSTANT** is permitted for the variable types VAR, VAR_EXTERNAL and VAR_GLOBAL. Constants, which are declared as global, must also be declared as constants when used externally.

The simultaneous use of **RETAIN** and **CONSTANT** makes no sense and is not permitted by IEC 61131-3 because constants must always be restored (statically) in the PLC after a power failure, thus making the RETAIN qualifier superfluous.

The qualifiers **R_EDGE** and **F_EDGE** indicate Boolean variables which can recognise rising or falling edges. They are permitted by IEC 61131-3 exclusively for the variable type VAR_INPUT. However, it is also conceivable to extend this mechanism to the variable types VAR and VAR_GLOBAL. Edge detection is performed implicitly by standard function blocks of IEC 61131-3 and is described in detail in Section 5.2.

```
VAR_OUTPUT RETAIN
  RetentiveVar   : BYTE;
END_VAR
VAR CONSTANT
  ConstantNum   : BYTE := 16#FF;
END_VAR
VAR_INPUT
  FallingEdge    : BOOL F_EDGE;
END_VAR
VAR_ACCESS
  LineEmpty      : CPU_LINE.%IX1.0    : BOOL READ_ONLY;
END_VAR
```

Example 3.17. Examples of the use of attributes/qualifiers with different variable types (shown in bold type)

The attributes **READ_ONLY** and **READ_WRITE** are reserved exclusively for the variable type VAR_ACCESS, (see also Chapter 6). No other qualifiers are permitted for VAR_ACCESS at the configuration level.

3.5.5 Graphical representation of variable declarations

The declaration part of POUs - as well as the code part - can be programmed both in graphical and in textual form. The graphical representation is used for better visualisation of the call interface and of the POU return values with their input and output variables.

The graphical facilities defined by IEC 61131-3 are adequate for simple variable declarations of the POU interface, but for declaration of arrays, retentive variables or initial values the textual representation must be used.

Table 3.14 shows which variable types and attributes of a POU can be represented (and therefore also declared) graphically as well as textually.

	Graphical representation
Variable types	
VAR	no
VAR_INPUT	yes
VAR_IN_OUT	yes
VAR_OUTPUT	yes
VAR_EXTERNAL	no
VAR_GLOBAL	no
VAR_ACCESS	no
Attributes with variable types	
RETAIN, NON_RETAIN	no
CONSTANT	no
R_EDGE, F_EDGE	yes[a]
READ_ONLY, READ_WRITE	no

a   Planned only for variable type VAR_INPUT

Table 3.14. Graphical representation for variable types and their attributes

Example 3.18 shows a declaration part in graphical representation with some of the possibilities specified in Table 3.14, as well as the textual version.

The graphical representation of the qualifiers RETAIN and CONSTANT and the other attributes is not dealt with in IEC 61131-3. This could, however, be implemented additionally for a programming system. Graphical representation of all variable types is also conceivable.

As illustrated in Example 3.18, the graphical representation of declarations visualises the calling interface (formal parameters) and return values of POUs.

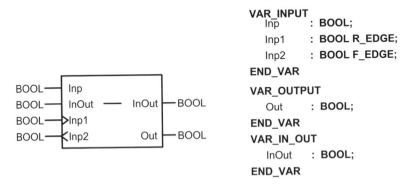

Example 3.18. Graphical and textual representation of the declaration part. Edge detection is symbolised by ">" and "<", VAR_IN_OUT with a continuous line.

The call of a POU can also be represented graphically. In this case, the actual parameters are assigned to the formal parameters and the return values are processed further.

This results in a uniform graphical view of the declaration and invocation of a POU, as shown for function block IO_Exam in Example 3.19.

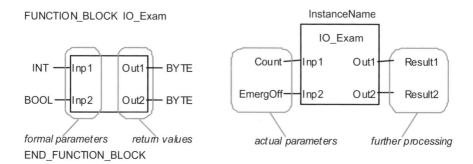

Example 3.19. Formal parameters and return values in the FB declaration (left) and supply of actual parameters in the FB invocation as well as further processing of the results (right)

For an explanation of formal parameters and return values see also Section 2.3.2.

4 The Programming Languages of IEC 61131-3

IEC 61131-3 provides three *textual languages* and three *graphical languages* for writing application programs. The textual languages are:

- Instruction list IL,
- Structured Text ST,
- Sequential Function Chart (textual version).

The graphical languages are:

- Ladder Diagram LD,
- Function Block Diagram FBD,
- Sequential Function Chart SFC (graphical version).

The code part of the textual language ST consists of a sequence of **statements** and the code part of the textual language IL of a sequence of **instructions**.

Statements in ST, a high-level language, consist of a combination of ST keywords which control the program execution and so-called *expressions*. Expressions, consisting of operators/ functions and operands, are evaluated at run time.

   In IL, an *instruction* consists of an *operator* or a *function* plus a number of *operands* (parameters). An operator usually has one (or no) operand and a function may have one or more (or no) parameters.

The graphical languages use graphic elements to formulate the desired behaviour of the PLC. Connecting lines or so-called *connectors* indicate the data flow between functions and function blocks.

In the following sections, the fundamental structure of the language or graphics constructs of each language are discussed; then the individual constructs are described in detail. Finally, an example is presented for every language.

K.-H. John, M. Tiegelkamp, *IEC 61131-3: Programming Industrial Automation Systems*, 2nd ed., DOI 10.1007/978-3-642-12015-2_4,
© Springer-Verlag Berlin Heidelberg 2010

4.1 Instruction List IL

Instruction List IL is a convenient assembler-like programming language. IL is universally usable and is often employed as a common *intermediate language* to which the other textual and graphical languages are translated.

4.1.1 Instruction in IL

IL is a line-oriented language. An *instruction*, which is an executable command for the PLC, is described in exactly one line. Empty instructions in the form of blank lines are also allowed.

A statement in IL consists of the elements represented in Figure 4.1.

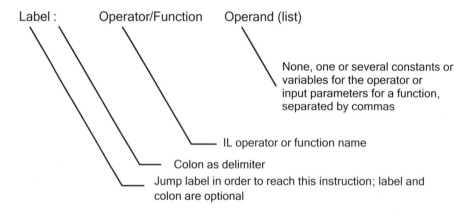

Figure 4.1. Instruction in IL; at least one blank space is required between operator and operand to distinguish these parts. **Comments** with (* ... *) brackets are allowed wherever a blank space may be inserted.

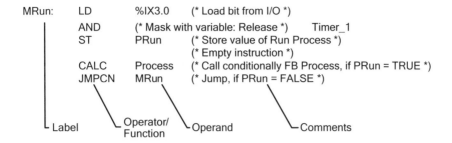

Example 4.1. Various IL instructions

The label and the comment of an IL line are optional. As of Edition 2 [IEC 61131-3], comments are allowed not only at the end of a line, but wherever a blank space may be inserted, as in the other languages. For ease of reading, however, the former convention of inserting comments only at the end of a line is continued to be applied in this book.

Labels are necessary to enable jumps in program execution to be performed from lines elsewhere in the program. A label on an otherwise empty line is allowed; the subsequent lines will be executed.

The individual *operators/functions* describe the desired operation; see Sections 4.1.3 and 4.1.4. The *operands* and input parameters are explained in Chapter 3 and Appendices A and B.
Comments are identical in all languages and are delimited by a pair of asterisks and brackets (*   *). This construct is used for informal comments on the contents of a line.

There is no defined format (such as column number) for the operators or operands; both parts can be separated by any number of blank spaces or tabs (see Example 4.1). The first character of an operator can start in any column.

Note: The semicolon ";" is not allowed in IL, either as a "start of comment" character (as used in many assemblers) or as a statement terminator (as used in Structured Text).

4.1.2 The universal accumulator (Current Result)

Standard assemblers are usually based on a real (hardware) accumulator of a processor, i.e. a value is loaded into the accumulator, further values are added, subtracted, ... and the result of the accumulator may finally be stored in a memory location.

IL also offers an *accumulator* called the *"Current Result"* (CR) . However, the *CR* does not have a fixed number of bits like real hardware accumulators. The IL compiler ensures that a virtual accumulator (including accumulator stack) of any storage width is always available. The number of bits depends on the data type of the operand being processed. The data type associated with the CR also changes to match the data type of the most recent operand.

Unusually for an assembler, IL has no specific processor status bits. The evaluation of a comparison operation produces a Boolean 0 (FALSE) or 1 (TRUE) result in the CR. Subsequent conditional jumps and calls use the TRUE or FALSE content of the CR as the condition during execution of the conditional jump or call.

```
VAR
    FirstOperand, SecondOperand, Result: INT := 10;
    StringOp: String[30] := '12345678901234567890';
    StringRes: String[25];
END_VAR
...
B1:    LD       FirstOperand     (* 10 {INT} *)
       ADD      SecondOperand    (* 20 {INT} *)
       ST       Result           (* 20 {INT} *)
       GT       0                (* TRUE, because 20 > 0 {BOOL} *)
       JMPC     B2
       (* jump, because CR=TRUE; present value of CR remains {BOOL} *)
       JMP      FarAway (* CR is not defined or present value: *)
       (* implementation-dependent *)
B2:    LD       StringOp         (* 12345678901234567890 {STRING} *)
       ST       StringRes        (* 12345678901234567890 {STRING} *)
(* .. *)

FarAway: (* ... *)
```

Example 4.2. The universal accumulator of IL. The value of the "Current Result" (CR) and its current data type (in {} brackets) after execution of the instruction are shown in the instruction comments.

A CR (Current Result) can be of type:

- Elementary data type,
- Derived data type (structure, array, etc.),
- Function block type.

The data capacity of the Current Result (the number of bits) is unimportant, as is shown by the character string operation in Example 4.2.

IL demands that two consecutive operations must be compatible; i.e. the data type of the current CR must be the same as the subsequent instruction data type.

In Example 4.3, an ADD operator with an operand of type INT requires the same data type in the CR (from the preceding instruction) to perform correctly.

```
VAR_INPUT
  FirstOperand: INT;
END_VAR
VAR
  SecondOperand, ThirdOperand: INT := 10;
  WordVar:        WORD;
END_VAR
  LD    1                   (* 1 {INT} *)
  ADD  FirstOperand         (* 11 {INT} *)
  ST    SecondOperand       (* 11 {INT} *)
  LT    ThirdOperand        (* FALSE {BOOL} *)
  AND  WordVar              (* Error: WordVar is of type WORD, not BOOL as CR *)
                            (* Data type mismatch *)
  ST  Exam43
```

Example 4.3. IL program containing an error: The data type of the CR (BOOL) does not match the data type of the "AND WordVar" instruction, which requires type WORD.

Some operators change the type of the CR, as can be seen from Examples 4.2 and 4.3.

Influence of the group of operators on the CR	Abbrev.	Example operators
Create	C	LD
Process	P	GT
Leave unchanged	U	ST; JMPC
Set to undefined	-	CAL = Unconditional call of a function block. (The following instruction has to load the CR anew, because on return from the FB the CR has no defined value.)

Table 4.1. Modification of the CR by different **groups of operators**

The list in Table 4.1 shows the different types of influence an operator can have on the CR. We will refer to this table later.

Leave unchanged means that an instruction passes the CR of the preceding instruction to the following one, without a change of value and type.

Set to undefined means that the subsequent instruction cannot use the CR. The first instruction in an FB (called via CAL) must therefore be a load (LD), jump (JMP), FB call (CAL) or return (RET) because these operations do not require a valid CR.

The IEC 61131-3 itself does not define *operator groups*; however, the groups help to explain more clearly the rules that are needed to combine the IL instructions successfully. The IEC describes the influence and evaluation of the CR only in a rudimentary manner. In the case of operations like AND and OR, the type and value of the CR before and after the operation is well defined. However, the standard does not discuss the validity of the CR value and data type after an unconditional jump. The operator groups referred to in the subsequent tables should therefore be seen as an aid in understanding the standard; they are an interpretation of the standard, but not part of it. They can be implemented differently in different programming systems.

4.1.3 Operators

This section describes all the *operators* defined for IL. Some operators have extensions called modifiers. An operator combined with a modifier like **N**, **(** or **C** has an extended meaning.

Modifiers in detail:

- N   Negation of operand,
- (Nesting levels by parenthesis,
- C   Conditional execution of operator (if CR=TRUE).

The functionality of these modifiers is explained by means of the subsequent examples.

Negation of the operand.
The operand is negated before carrying out the instruction:

```
VAR     Var1: BOOL := FALSE;    END_VAR
LDN     FALSE          (* TRUE {BOOL} ; equivalent to LD 1 *)
ANDN    Var1           (* TRUE {BOOL} *)
```

Example 4.4. Negation of the operand by the modifier N. The value and type {} of the current result CR are shown in the comment after the instruction.

Nesting levels by parenthesis.

It is possible to perform a logic operation on the CR and the result of a whole sequence of instructions by using the *parenthesis modifier*. When the "("modifier is detected, the operator type and the current CR value and data type are "saved" and a new value and type are loaded in the CR. When the closing parenthesis ")" is detected, the deferred value and type are "retrieved" and are operated on using the modified operator and the current CR value. The result is stored in the CR.

The standard does not describe how the deferred CR is saved and restored but typically a system stack is used for storing temporary data.

```
    LD      Var1      (* Load value of Var1 in CR and then defer it *)
┌   AND(    Var2      (* Calculate the expression in parentheses and *)
│   OR      Var3      (* associate the result with the Current Result (Var1) *)
└   )
    ST      Var4      (* Store the result  *)
```

Example 4.5. Blockwise computation of parenthesis operators. The sequence calculates the Boolean expression Var4 := Var1 $\wedge$ (Var2 $\vee$ Var3).

It is possible to program several levels of these parenthesised blocks:

```
LD    1             (* 1 *)
ADD(  2             (* 2 *)
    ADD(    3       (* 3 *)
        ADD   4     (* 7 *)
    )               (* 9 *)
)                   (* 10 *)
ST    Var1          (* 10 *)
```

Example 4.6. Computation of nested expressions in parentheses. The comment shows the value (result) of the Current Result after the operation.

Note: The format of columns shown in Example 4.6 ("tabs") is not mandatory. However, the ability to document nested expressions using indents increases the readability and therefore the quality of the program.

The Standard allows 2 ways of writing parenthesis operators; see Table 4.2

a) Short form of parenthesis operator	b) Long form of parenthesis operator

LD   1
ADD(2
   ...
)
	LD 1
	ADD(
	LD 2
	   ...
	)

Table 4.2. Short form (a) and long form (b) of parenthesis operators

Note: It is not only the value of the CR which can be deferred using the parenthesis modifier, the data type can also be deferred so long as data type consistency is maintained throughout the entire instruction sequence.

Conditional execution of operators.
Several operators, like GT, generate a Boolean value stored in the Current Result. If the Boolean value is TRUE, the following (conditional) instruction (like a conditional jump) is executed. Otherwise, the instruction following the conditional instruction is given the control of the processor:

```
LD      FirstOperand
GT      20              (* Generate CR = TRUE if CR > 20, else FALSE *)
JMPC  B2                (* Jump to B2, if CR = TRUE, else next line *)
JMP    FarAway          (* Jump to FarAway, independent of the CR *)
```

Example 4.7. Conditional and Unconditional form of the jump JMP/JMPC

The next three tables give a complete overview of operators and modifiers.
  Some operators are reserved for Boolean data types and/or Bit String (ANY_BIT), some for generic data types, others for jumps and calls. The type compatibility of any standard function of the same name (such as ADD, ...) in IEC 61131-3 (see Appendix A) must be ensured.

IEC 61131-3 demands Boolean operands or Bit String (ANY_BIT) for the operators of Table 4.3.

BOOL & Bit String:

Operators		Group of operators [a]	Description
LD	LDN	C	Load operand (negated) to CR
AND	ANDN	P	Boolean AND of the (negated) operand
AND (	ANDN (		with the CR
OR	ORN	P	Boolean OR of the (negated) operand with
OR (	ORN (		the CR
XOR	XORN	P	Boolean Exclusive-OR of the (negated)
XOR (	XORN (		operand with the CR
NOT		P	Bit-wise Boolean negation (unit complement)
ST	STN	U	Store CR to the operand
S		U	Set operand TRUE, if CR=1
R		U	Set operand FALSE if CR=1
)		P	Closing parenthesis; evaluate deferred operation

a   Description see Table 4.1.

Table 4.3. Operators with Boolean and Bit String operands (Type ANY_BIT)

Why do we need the definition of the "parenthesis" modifier? IEC 61131-3 defines standard functions like AND that allow data type ANY_BIT und thus adequately extend the operator functionality. However, in IL, it is not possible to include a computation result in a function as an operand as is allowed in ST or FBD. Therefore the parenthesis modifier was defined to make such computations possible.

```
LD    V0
AND  V1 OR V2 (* Error *)
ST    V3
(* Attention: result not *)
(* identical to *)
(* LD    V0 *)
(* AND V1 *)
(* OR    V2 *)
(* ST    V3 *)
```

```
LD    V0
AND(  V1
OR    V2
)
ST    V3
```

```
V3 := V0 AND
     (V1 OR V2);
```

| a) Call by std. function/ operator | b) Operator | c) ST | d) FBD |

Example 4.8. This shows how complex expressions of logical operators/std. functions are handled as a sequence of instructions using the parenthesis modifier.

The operators S and R are abbreviations for conditional storing. The operator S stands for "Store Conditional STC", R is the notation for "Store Conditional Not STCN".

Operators		Group of operators [a]	Description
LD		C	Load operand to CR
ST		U	Store CR to operand
ADD	ADD (	P	Add operand to CR
SUB	SUB (	P	Subtract operand from CR
MUL	MUL (	P	Multiply operand with CR
DIV	DIV (	P	Divide CR by operand
MOD	MOD(	P	Modulo division
GT	GT (	P	CR > operand (greater than)
GE	GE (	P	CR >= operand (greater than or equal)
EQ	EQ (	P	CR = operand (is equal)
NE	NE (	P	CR <> operand (is not equal)
LE	LE (	P	CR <= operand (less than or equal)
LT	LT (	P	CR < operand (less than)
)		P	Closing parenthesis level

a   Explanation see Table 4.1.

Table 4.4. Operators for operands of generic numeric data type (type ANY_NUM)

Table 4.4 also includes the operators LD and ST. Both can be used not only for Boolean data types but also for all data types. The property that these operators support data of type ANY is similar to the functionality of "overloading of functions". The data type of the variable/constant operand defines the operator data type. It is the compiler's responsibility to activate the correct operator for the given data type.

Operators comparing two values (CR and operand) generate TRUE in the CR when the condition is TRUE, otherwise FALSE.

Jump or Call:

Operators	Group of operators [a]	Description
JMP	- or U[b]	(Un)conditional jump to a jump label
JMPC JMPCN	U	
CAL	-	(Un)conditional call of a function block
CALC CALCN	U [c]	
RET	- or U[d]	(Un)conditional return from a function or function block
RETC RETCN	U	
Function name	P[c]	Function call

A   Explanation see Table 4.1.
B   Implementation-dependent
C   Valid for following operator (but undefined at start of function or FB execution!)
D   For functions: Function value ("U")

Table 4.5. Operators for Jump and Call

The operand of a jump operator is a label. The operand of an FB call (CAL, CALC, CALCN) is the name of the FB instance.

4.1.4 Using functions and function blocks

Calling a function.
A *function* is called in IL by simply writing the name of the function. The actual parameters follow separated by commas. The syntax is equivalent to that of an operator with several operands.

Formal parameters can also be assigned actual values line by line using ":=". However, this does not apply to the operators valid in IL, where only the above alternative of value assignment is allowed.

The first parameter of a function is the Current Result (CR). Therefore, this value must be loaded into the CR just before the function is called. The first operand, which is used in the function call, is actually the second parameter of the function and so on, see Example 4.9.

Actual parameters	With formal parameters	With formal parameters

LD   1
LIMIT   2, 3
	LIMIT(
	    MN := 1,
	    IN := 2,
	    MX := 3
	)
		LIMIT(
		    MX := 3,
		    IN :=2,
		    MN :=1
		)

Example 4.9. Equivalent calls of a function using actual parameters and formal parameters, latter example in an arbitrary sequence

A function returns at least one output value. This is stored in the CR and has the data type specified for the function. Other output parameters of the function are returned through parameter assignments, see Section 2.5 and Example 4.10. When a function is called without formal parameters, the sequence of declarations must be adhered to. In the case of formal parameter assignments, this is done line by line with a concluding parenthesis.

```
LIMIT(
        EN:= ErrorCond,
        MN :=1,
        IN :=2,
        MX := 3,
        ENO => ErrorCondLim       (* => indicates: ENO is output parameter *)
        )
```

Example 4.10. Function call equivalent to that in example 4.9 with formal parameters with execution control EN/ENO and output parameter ENO. The result of LIMIT is stored in the CR.

The programming system assigns the function value to a variable with the function name; this name is declared automatically and does not have to be specified separately by the user in the declaration part of the calling function block.

The parameters EN/ENO are used according to the functionality described in Section 2.7.1. A POU with EN/ENO parameters must be called with formal parameters, and not only with actual parameters.

Function call:	*Definition of the function*

```
VAR
   FirstFunPar: INT :=10;
   Par2: INT := 20;
   Par3: INT := 30;
   Sum: INT;
END_VAR

LD            FirstFunPar
UserFun       Par2, Par3
(* Second call: *)
UserFun       Par2, Par3
ST            Sum
```

```
FUNCTION      UserFun : INT
   VAR _INPUT
      FunPar1, FunPar2 , FunPar3: INT;
   END_VAR

   LD     FunPar1
   ADD    FunPar2
   ADD    FunPar3
   ST     UserFun  (* Ret value,can be omitted *)

   RET              (* Can also be omitted*)
END_FUNCTION
```

Example 4.11. Two calls of the user function, UserFun. At the first call, FunPar1 has the value 10, at the second, 60. FunPar2 is 20 and FunPar3 is 30. At the end 110 is stored in Sum.

Calling a function block.

An FB can be activated by the operator CAL (or CALC for conditional and CALCN for conditional negated calls). IEC 61131-3 describes three methods of passing parameters to an FB in the IL language:

1) Using a call including a list of actual input and output parameters in brackets,
2) Loading and saving the input parameters before calling the FB,
3) Calling "implicitly" by using the input parameters as operators.

Method 3 is only valid for standard FBs, not for user FBs. This manner of activating FBs is seldom used in practice.

These three call methods are demonstrated in the following example.

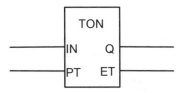

Figure 4.2. Standard function block TON with input and output parameters

```
VAR
   Rel, Out:        BOOL := 0; (* Release - Input; Output *)
   Time1:           TON;       (* Std - FB TON has the formal parameters *)
                               (* IN, PT (input) and Q, ET (output) *)
   Value            TIME;      (* Set - Input *)
END_VAR
```

(* Method 1 *)	(* Method 2 *)		(* Method 3 *)	
(* Supplying the parameters: *)				
	LD	t#500ms	LD	t#500ms
	ST	Time1.PT	PT	Time1
	LD	Rel	LD	Rel
	ST	Time1.IN		

(* Call: *)
CAL Time1 (
    IN:=Rel,
    PT:= t#500ms,
    Q=>Out, (* Output p. 1 *)
    ET=>VALUE (* Output p. 2 *)
)
	CAL	Time1	IN	Time1

	(* Utilisation of the output parameters *)
	    LD Time1.Q
	    ST Out
	    LD Time1.ET
	    ST Value			

Example 4.12. Time1 declared as an instance of FB TON, the loading of the input parameters, the call of the instance and the evaluation of the output parameters.

The declaration part and the evaluation of the output parameters are identical for all three methods. The methods differ only in the supplying of the input parameters and the FB call; see also Section 2.8.4.

If a programming system supports several methods, the supply, call and evaluation methods can be mixed.

In modification of method 1, a call can also be made with actual parameters, as in a function call. In this case, the actual parameter values are written in their defined sequence, separated by commas.

Using method 3, the PT operator initialises the PRESET time. The IN operator starts the FB Time1. Method 3 will not change input parameters in a command line that are not addressed, i.e. the values from previous assignments will be used. Method 3 should be used with care because it is not always clear when calling a function block whether an input parameter is defined or whether the function block is executed.

4.1.5 IL example: Mountain railway

The following example is a simplification of a control system programmed for a *mountain railway*.

It is written here in IL. The graphical representation in LD is shown in Section 4.4.5

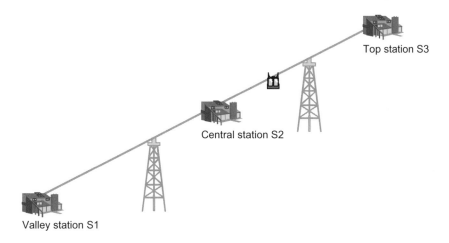

Figure 4.3. Example of a mountain railway with three stations and one cabin.

The program has to control the following features:

1) The sensors S1, S2, S3 send TRUE (1) when the cabin arrives at one of the stations. The counter StationStop stores the total number of stops at stations.
2) The motor to move the cabin is controlled by the following signals:
   * Direction: Forward (TRUE) / Backward (FALSE) (retained variables)
   * StartStop: Start (TRUE), Stop (FALSE)
3) Inside the cabin, a switch DoorOpenSignal is used to open/close the door. DoorOpenSignal equal to TRUE signals "Open the door", FALSE "Close the door".
4) The motor to move the door has two actuators: OpenDoor and CloseDoor. Both actuators are triggered by a rising edge of one of the two signals.
5) A button MRStart starts the whole system. The button MREnd is used to shut down the system.

6) A danger signal must be activated between shutdown of the railway and a new start.

A detailed description of this control program can be found in Section 4.4.5.

```
FUNCTION_BLOCK MRControl
  VAR_INPUT
    MRStart:        BOOL R_EDGE;    (* Edge-triggered button to start the railway *)
    MREnd:          BOOL;           (* Switch for initiation of the end of operation *)
    S1, S2, S3:     BOOL R_EDGE;    (* Edge-triggered sensors in every station *)
    DoorOpenSignal:BOOL;            (* Switch to open the door. 1: open; 0: close! *)
  END_VAR

  VAR_IN_OUT
    StartStop:      BOOL;           (* Cabin moving: 1; not moving: 0 *)
  END_VAR

  VAR_OUTPUT
    OpenDoor:       BOOL;           (* Motor to open the door *)
    CloseDoor:      BOOL;           (* Motor to close the door *)
  END_VAR

  VAR_OUTPUT RETAIN
    EndSignal:      BOOL;           (* Warning signal for Power Off (retained) *)
  END_VAR

  VAR
    StationStop:    CTU;            (* Standard FB (counter) for cabin stops *)
    DoorTime:       TON;            (* Standard FB (delayed start) of cabin *)
  END_VAR

  VAR RETAIN
    Direction:      BOOL;           (* Current direction up or down *)
  END_VAR

  (* System running for the first time after power on? Yes: Reset the end signal *)
  (* activated by the last shutdown *)
  LD      MRStart                   (* The first call? *)
  R       EndSignal                 (* Yes: Reset the warning signal *)
  JMPC    ResCount

  (* Not the first call! *)
  JMP     Arrive

  (* Reset the station counter *)
  ResCount: LD      1
  ST      StationStop.RESET         (* Reset the counter *)
  LD      9999
  ST      StationStop.PV            (* Maximum *)
  CAL     StationStop               (* Call of the FB instance StationStop *)
  JMP     CloseCabin

  (* Increase the counter StationStop when the cabin arrives at a station*)
  Arrive: LD        S1              (* Sensors are edge-triggered! *)
  OR      S2
  OR      S3
```

```
(* Stop if the CR is now TRUE *)
R         StartStop                     (* Stop cabin *)
CALC      StationStop (RESET:= 0,CU:=1) (* If stop-> increase counter *)

(* Change of direction? *)
LD        S1
XOR       S3
JMPCN     NoDirChange                   (* either S1 or S3 are TRUE? *)

(* Change direction of cabin *)
LD        Direction
STN       Direction

(* Condition to open cabin: cabin stops and door open switch is activated *)
NoDirChange: LD          DoorOpenSignal
ANDN      StartStop
ST        OpenDoor

(* End signal for railway and cabin in a station -> POU end *)
LD        MREnd          (* Power-off activated? *)
ANDN      StartStop
S         EndSignal
JMPC      PouEnd

(* Door switch signals to shut the door *)
CloseCabin: LD           DoorOpenSignal
STN       CloseDoor

(* Cabin start 10 seconds after activation of the door switch *)
LDN       DoorOpenSignal
ANDN      StartStop
ST        DoorTime.IN
LD        T#10s
ST        DoorTime.PT
CAL       DoorTime
LD        DoorTime.Q                    (* Time finished? *)
S         StartStop

RET (* Return to the calling POU *)
PouEnd:
END_FUNCTION_BLOCK
```

Example 4.13. Control of a mountain railway by FB MRControl written in IL. Please refer to the CD for an optional version of the calling PROGRAM (MRMain.POE).

4.2 Structured Text ST

Like IL, the programming language *Structured Text ST* is a textual language of IEC 61131-3. ST is called a High-Level Language, because it does not use low-level, machine-oriented operators but offers a large range of abstract statements describing complex functionality in a very compressed way.

Advantages of ST (compared to IL):

- Very compressed formulation of the programming task,
- Clear construction of the program in statement blocks,
- Powerful constructs to control the command flow.

These advantages also bring their own disadvantages:

- The translation to machine code cannot directly be influenced by the user since it is performed automatically by means of a compiler.
- The high degree of abstraction can lead to a loss of efficiency (compiled programs are in general longer and slower).

A ST algorithm is divided into several steps (statements). A **statement** is used to compute and assign values, to control the command flow and to call or leave a POU.

PASCAL and C are comparable high-level programming languages in the PC world.

4.2.1 ST statements

A ST program consists of a number of *statements*. Statements are separated by semicolons (;). Unlike IL, the "end of line", line feed, is interpreted syntactically as a space. This means that a statement can be extended over several lines or several statements can be written on a single line.

Comments are framed by parentheses and asterisks "(* comment *)" and may be used within statements in all places where spaces are allowed:

A := B (* elongation *) + C (* temperature *);

Although the language is in free-format, the user is strongly recommended to use the layout and comment features as a documentation aid to improve readability.

The statements of ST are summarised in Table 4.6.

Keyword	Description	Example	Explanation
:=	Assignment	**d := 10;**	Assignment of a calculated value on the right to the identifier on the left.
	Call of an FB [a]	**FBName(** **Par1:=10,** **Par2:=20,** **Par3:=>Res);**	Call of another POU of type FB including its parameters. ":=" for input parameters, "=>" for output parameters
RETURN	Return	**RETURN;**	Leave the current POU and return to the calling POU
IF	Selection	**IF** d < e **THEN** f:=1; **ELSIF** d=e **THEN** f:=2; **ELSE** f:= 3; **END_IF;**	Selection of alternatives by means of Boolean expressions.
CASE	Multi-selection	**CASE** f **OF** 1: g:=11; 2: g:=12; **ELSE** g:=FunName(); **END_CASE;**	Selection of a statement block, depending on the value of the expression "f".

FOR	Iteration (1)	**FOR** h:=1 **TO** 10 **BY** 2 **DO**
		f[h/2] := h;
		END_FOR;
			Multiple loop of a statement block with a start and end condition and an increment value.

WHILE	Iteration (2)	**WHILE** m > 1 **DO**
		n := n / 2;
		END_WHILE;
			Multiple loop of a statement block with end condition at the beginning.

REPEAT	Iteration (3)	**REPEAT**
		i := i*j;
		UNTIL i < 10000
		END_REPEAT;
			Multiple loop of a statement block with end condition at the end.

EXIT	End of loop	**EXIT;**	Premature termination of an iteration statement
;	Dummy statement	**; ;**	

a   Function calls are not statements in themselves but can be used within an expression as operands

Table 4.6. ST statements

ST does not include a jump instruction (GOTO). This is not a functional restriction, especially since jump instructions encourage the use of non-structured programs. All conditional jumps can also be programmed via an IF structure. Due to the specific PLC environment (such as its real-time requirement), however, jumps may increase the efficiency to quickly leave complex nested statements, for example, after an error condition has been found.

As shown in Table 4.6, every statement uses either the value of individual variables (h, e), constants (10) or the result of a computation of several variables (e.g. multiplication: i * j).

The part of a statement that combines several variables and/or function calls to produce a value is called an *expression*.

This basic element of ST is described in the following section. Understanding expressions is essential to the understanding of how ST statements work.

4.2.2 Expression: Partial statement in ST

Expressions produce the **values** necessary for the processing of statements. An expression value can be generated by an individual operand or by the result of logic operations on several operands.

Examples of expressions (see Table 4.6):
10   n / 2   i < 10000   FunName()

An expression consists of *operands* and associated ST *operators*. Its fundamental construction is shown in the following diagram:

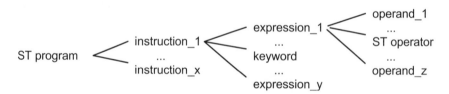

Figure 4.4. An ST program is constructed by using statements. These consist of keywords and expressions. Expressions consist of operands and ST operators.

Operands.
Operands may consist of the following :

- **Literal** (numeric, alphanumeric characters, time)
  Examples: 10, 'abcdef', t#3d_7h,
- **Variables** (single-/ multi-element variables)
  Examples: Var1, Var[1,2], Var1.Substruc,
- **Function call**; more precisely: the **function return value**
  Example: FunName(Par1, Par2, Par3),
- Another expression
  Example: 10 + 20.

Operators.

For logic operations on the operands within expressions, the following *operators* are defined in ST :

Operator	Description	Example and value of the expressions	Priority
( )	Parenthesis	(2 * 3) + (4 * 5) Value: 26	high
	Function call	**CONCAT** ('AB', 'CD') Value: 'ABCD'	
-	Negation	-10 Value: -10	
NOT	Complement	**NOT** TRUE Value: FALSE	
**	Exponentiation	3 ** 4 Value: 81	
*	Multiplication	10 * 20 Value: 200	
/	Division	20 / 10 Value: 2	
MOD	Modulo	17 **MOD** 10 Value: 7	
+	Addition	1.4 **+** 2.5 Value: 3.9	
-	Subtraction	3 - 2 - 1 Value: 0	
<, > , <=, >=	Comparison	10 > 20 Value: FALSE	
=	Equality	T#26h **=** T#1d2h Value: TRUE	
<>	Inequality	8#15 **<>** 13 Value: FALSE	
&, AND	Boolean AND	TRUE **AND** FALSE Value: FALSE	
XOR	Boolean Exclusive OR	TRUE **XOR** FALSE Value: TRUE	
OR	Boolean OR	TRUE **OR** FALSE Value: TRUE	low

Table 4.7. Operators of ST as elements of expressions incl. precedence rule

Unlike a function call, which is an expression, an FB call is a statement. It has no return value and, consequently, FB calls are **not** allowed within an expression.

If there are equivalent standard functions (such as exponentiation, multiplication, etc.) to the operators mentioned above, they have to obey the rules concerning the data types of the input parameters and the value/type of the function return value:

For example, the following expressions are equivalent:

"x + y" and the standard function **ADD** (x, y).

If an expression consists of multiple operators, precedence rules are used to determine the sequence in which the expression parts are evaluated. See the explanation of the following example:

(115 - (15 + 20 * 2)) MOD 24 MOD 10 + 10

Example 4.14. Processing operators in order of precedence

1) In an expression, the operators are executed in the order of precedence shown in Table 4.7 :

Compute the innermost expression in parentheses first since parentheses have the highest priority of the operators:

Compute multiplication: (20 * 2 = 40)

Compute addition: (15 + 40 = 55)

This yields 55

Compute the next level of parenthesis:

Compute subtraction: (115 - 55 = 60)

This yields 60

Compute the modulo operator: (60 MOD 24 = 12)

2) If an expression contains several operators with the same precedence, they are processed from left to right.

Compute: 60 MOD 24 MOD 10 = 2

3) Operands of an operator are processed from left to right.

First, compute the expression to the left of the "+" sign i.e.:

Compute: (115 - (15 + 20 * 2)) MOD 24 MOD 10 (see 2)

This yields 2

Then, compute the expression to the right of the "+" sign:

Compute 10

Compute the addition operator

This yields 12

The data type of this value depends on the variable to the left of the ":=".
(in Example 4.14 this variable must have a data type of the generic class ANY_INT).

Parenthesis should not only be used to define a specific processing sequence, but also to increase the readability of a complex expression. Also, wrong assumptions about precedence can be avoided using parentheses.

Function as operator.

A *function* call consists of the function name and its parameter list separated by commas and enclosed in parentheses. Either a formal or an actual parameter list may be used. Since a *formal parameter* list includes both the name of each parameter (":=") (input parameter) and its actual value ("=>") (output parameter), the sequence of formal parameters is arbitrary. Parameters may be omitted and a default value will be provided. Actual parameter lists consist only of actual values separated by commas. In this case, all parameters must be present in their correct sequence i.e. the sequence in which the function parameters are defined in the function declaration.

```
ADD (1, 2, 3);                       (* Actual parameters without formal parameters*)
LIMIT (MN := 0, MX := 10, IN := 4);  (* Formal parameters, sequence irrelevant *)
USERFUN(OutP1:=10, OutPar=>Erg); (* User function with 1 input parameter,
                                      1 function value (return value)
                                      and 1 additional output parameter *)
```

Example 4.15. Call of the standard functions ADD with actual parameters, LIMIT with formal parameters and USERFUN as an example with additional output parameter

In addition to the standard functions, user-defined functions are also allowed. Unlike FB calls (statements), function calls are expressions. Functions return just one result (together with optional output parameter) and do not have side effects. A function always returns the same return value and output parameter values when the same input parameters are provided. The use of global variables within functions is forbidden. This is necessary to ensure that functions have no side effects.

4.2.3 Statement: Assignment

An *assignment* statement replaces the value of a single- or multi-element variable (on the left-hand side of the operator) by the value of the expression evaluated on the right-hand side of the equal sign.

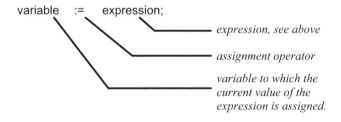

variable   :=   expression;

expression, see above

assignment operator

variable to which the current value of the expression is assigned.

Figure 4.5. Assignment in ST

```
TYPE MulVar: STRUCT Var1: INT; Var2: REAL; END_STRUCT; END_TYPE

VAR   d:   INT;
      e:   ARRAY [0..9] OF INT;
      f:   REAL;
      g:   MulVar (Var1:=10, Var2 := 2.3);
      h:   MulVar;
END_VAR

d     := 10;                    (* Assignment *)
e[0]  := d ** 2;  h := g;       (* Two assignments in one line *)
d     := REAL_TO_INT(f);        (* Assignment evaluating a function call *)
```

Example 4.16. Examples of statements of type assignment ":="

The first assignment of Example 4.16 stores the value 10 in the variable "d". The next statement assigns the value 100 to the first element of the array "e". The multi-element variable "h" is assigned all the values stored in "g" (10 for "Var1" and 2.3 for "Var2").

As in the second statement line of this example, it is possible to write several statements in one line. However, this should be avoided for reasons of clarity because shortness is no quality criterion for source code. Suitable structuring with lines and indenting (e.g. in complicated IF structures) can considerably increase the clarity and maintainability of the program.

When using assignments, it is important to remember data *type compatibility*. If the right and the left side of the assignment have different data types, a type conversion function is necessary to avoid a data type conflict.

Within a function body, an assignment statement assigning the result of an expression evaluation to the function name is mandatory. The data type of the function must match that of the expression result. Several such assignments are possible. The last such assignment executed before return from the function is the return value of the function.

```
FUNCTION Xyz: MulVar
    VAR_INPUT Factor1, Factor2:INT; END_VAR
    VAR_TEMP Tmp: MulVar; END_VAR


    Tmp.Var1 := 10 * Factor1;
    Tmp.Var2 := 4.5 * INT_TO_REAL(Factor2);
    Xyz := Tmp;
END_FUNCTION
```

Call of function Xyz:
```
...
VAR i: INT; z: MulVar; END_VAR

z := Xyz (20, 3);
i := z.Var1;
...
```

Example 4.17. Example of both function definition with return value assignment and call of the function (for the type definition of MulVar, see Example 4.16).

4.2.4 Statement: Call of function blocks

In ST, an *FB* is activated by its name and an argument list in parentheses. The arguments consist of the *formal parameter* names and the assignment of the actual values assigned via ":=". For an output parameter, assignment is made via "=>". The sequence of the parameter assignments is irrelevant. If parameters are omitted, the FB uses either the default initial value for that data type, if this is the first call to that FB, or the value assigned during the previous call. Calling via actual parameter is also possible in ST. As in IL, the actual values are passed on separated by commas.

```
FUNCTION_BLOCK FbType
    VAR_INPUT
        VarIn: INT, VarH: INT := 1;
    END_VAR
    VAR_OUTPUT VarOut: INT := 5;
    END_VAR

    IF (VarIn > VarH) THEN
        VarOut := 10;
    END_IF;
END_FUNCTION_BLOCK
```

Call of FB FbName:
```
...
VAR FbName: FbType; RES: INT;
END_VAR

FbName();
(* FbName.VarOut == 5 *)
FbName(VarIn := 3, VarOut=>Res);
(* Alternative call: FbName(3, 1, Res *);
(* FbName.VarOut == 10,
copied to RES *)
...
```

Example 4.18. Example of both an FB definition with value assignment to the output parameter and a call of the FB.

4.2.5 Statement: RETURN

The *RETURN* statement is used to leave a function, FB or program, even before completion.

Please keep in mind that in the case of a function, an assignment to a variable with the name of the function (*function value*) must take place before a return statement is executed. If the output parameters of an FB are not assigned values within the FB, they have the initial value for their data type (or the value stored during the previous write access).

```
...
(* if variable x less than y, exit POU *)
IF x < y THEN RETURN;
END_IF;
...
```

Example 4.19. An example of using RETURN to return to the calling POU before reaching the end of the called POU.

4.2.6 Statement: Selection and Multi- selection

These two statement types are used to execute specific statements depending on a Boolean condition.

In the following examples, "statements" represents zero or more *statements*; and "condition" represents Boolean expressions evaluating to the value TRUE or FALSE.

Selection.

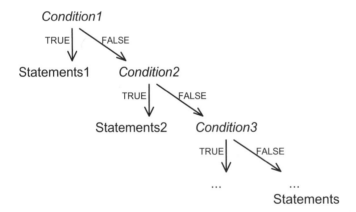

Figure 4.6. Binary tree of an IF statement, consisting of Boolean conditions and ST statements

Condition1 evaluates to the Boolean value TRUE or FALSE. If TRUE, "Statements1" is executed and the IF statement is finished. If FALSE, the next condition, "Condition2", is evaluated … and so on until the last block of statements, "Statements", is executed if all of the conditions are false. This is the "ELSE" part which ends the *IF selection statement.*

Nested IF statements are permitted.

IF expression **THEN** statement block;	*Execute the statement block if expression evaluates to* **TRUE**.
ELSIF expression **THEN** statement block;	*Evaluate expression if the previous expression is* **FALSE**. *Execute the statement block if the current expression is* **TRUE**. *It is possible to omit this part of an IF statement or to repeat it as many times as necessary.*
ELSE statement block;	*Execute the statement block if no previous expression evaluates to* **TRUE**. *It is possible to omit this part of an IF statement.*
END_IF;	*End of the IF statement (mandatory).*

Figure 4.7. IF selection. A statement block consists of zero or more statements. (ST keywords are in bold type).

The program for the selection tree in Figure 4.6 is shown below in Example 4.20:

```
IF Condition1 THEN
    Statements1;
    (* Execute Statements1, if Condition1 is TRUE, continue after " End_of_IF"; otherwise: *)
    ELSIF Condition2 THEN
        Statements2;
        (* Execute Statements2, if Condition2 is TRUE, continue after " End_of_IF"; otherwise:
        *)
        ELSIF Condition3 THEN
        ...
    ELSE Statements;
    (* Execute Statements if no previous condition evaluates to TRUE *)
END_IF;
(* End_of_IF *)
```

Example 4.20. Example of an IF cascade

Multi-selection.

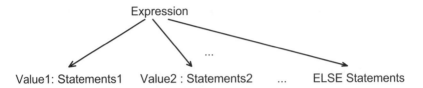

Figure 4.8. Switch with an expression of the type ANY_INT or enumeration, a list of values and corresponding statement blocks. A statement block is executed if the value corresponds to the evaluated expression.

After evaluating the expression, Value1, Value2… are compared with this expression. If one value fits, the statements of the corresponding statement block are executed. If the value of the expression is not found, the statement block of the ELSE (default value) branch is executed (it thus stands for all values that are not included in the list). If the expression does not produce any of the specified values and there is no default specified, no statement is executed. Values are specified in the form of an enumeration constant or one or more integer constants separated by commas. It is also possible to specify a range of integer constants separated by "..", e.g. 3..5 is equivalent to 3,4,5.

CASE expression **OF**	
case_value : statement block;	*Execution only if case_value corresponds to the evaluated expression. This part of the CASE statement can be written several times (with different case_values)*
	case_value is one ore more integer constants (SINT, INT, DINT, LINT) or a range of integer values: "value1..value2".
ELSE statement block;	*Execution only if no case_value corresponds to the expression. The ELSE part can be omitted.*
END_CASE;	*End of the CASE statement (mandatory)*

Figure 4.9. CASE statement. The statement block consists of zero or more statements (the ST keywords are in bold type).

The program for the multi-selection in Figure 4.8 is shown below in Example 4.21:

```
CASE VarInteger OF
    1:      Statements1;
            (* Execute Statements1, if VarInteger is TRUE, continue after " End_of_CASE";
                otherwise: *)
    2,3:    Statements2;
            (* Execute Statements2, if VarInteger is 2 or 3, continue after " End_of_CASE";
                otherwise: *)
    10..20: Statements3;
            (* Execute Statements3, if VarInteger is between 10 and 20, continue after
                "End_of_CASE"; otherwise: *)
  ELSE    Statements;
            (* Execute Statements if no comparison succeeded *)
END_CASE;
(* End_of_CASE *)

(* Continue *)
```

Example 4.21. Multi-selection with several statement blocks and a default "ELSE" block when no match is found.

4.2.7 Statement: Iteration

In order to execute a number of statements repeatedly, three language constructs are available in ST: WHILE, REPEAT and FOR.

WHILE and REPEAT statements.
A statement block is executed depending on the evaluation of a Boolean expression. The loop is active as long as the expression evaluates to TRUE (WHILE) or FALSE (REPEAT). WHILE tests the expression before the execution of the statements, whereas REPEAT tests it afterwards. With the REPEAT loop, the statements are executed at least once.

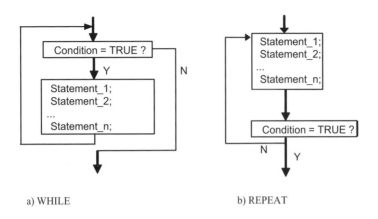

a) WHILE   b) REPEAT

Figure 4.10. Repeated execution of statements using WHILE (a) or REPEAT (b). Condition is a Boolean expression.

WHILE expression **DO**
      statement block;   *Loop the statement block as long as expression is evaluated to TRUE.*
END_WHILE;   *End of the WHILE statement*

Figure 4.11. WHILE statement. Statement block consists of zero or more statements (keywords are in bold type).

REPEAT
      statement block;
  UNTIL expression   *Repeat the statement block as long as expression is evaluated to FALSE.*
END_REPEAT;   *End of the REPEAT statement*

Figure 4.12. REPEAT-statement. Statement block includes zero or more statements (the ST keywords are in bold type).

In Example 4.22, an integer array is scanned to look for the biggest value in the array. This is assigned to a variable named Maximum.

```
VAR  TestField: ARRAY [1..5] OF INT := [2, 16, 4, 7, 32];
        Index: INT := 1; IndexMax: INT := 5; Maximum: INT := 0;
END_VAR
              ...
```

WHILE **Index <= IndexMax** DO	REPEAT

IF TestField[Index] > Maximum THEN
        Maximum := TestField[Index];
END_IF;
Index := Index + 1;
	IF TestField[Index] > Maximum THEN
	        Maximum := TestField[Index];
	END_IF;
	Index := Index + 1;

END_WHILE;	UNTIL **Index > IndexMax** END_REPEAT;

...

Example 4.22. Statement loop (within the rectangle) using a REPEAT or WHILE statement.

FOR statement.

The third type of repeat instruction includes a special integer control variable. This variable is initialised by default or with a user-defined value. At the end of a loop cycle the value of the control variable is incremented (or decremented) by a predefined increment value. The statement block is executed until the value of the control variable exceeds the final value in either the positive or the negative direction (depending on whether "increment" is positive or negative).

The statement block is only executed if the value of the control variable is within the initial/end interval. The initial, end and increment values are integer expressions.

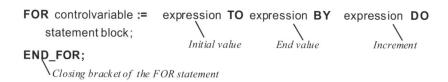

Figure 4.13. FOR statement. Initial, end and increment values of the control variable are determined by evaluating the relevant expression. The statement block consists of zero or more statements (ST keywords are in bold type).

See the next example showing the program for Example 4.22 written using a FOR statement:

```
VAR    TestField: ARRAY [1..5] OF INT := [2, 16, 4, 7, 32];
       Index: INT; IndexMax: INT := 5; Maximum: INT := 0; END_VAR
...
FOR Index := 1 TO IndexMax BY 1 DO
   IF TestField[Index] > Maximum THEN
       Maximum := TestField[Index];
   END_IF;
END_FOR;
...
```

Example 4.23. Repeated processing of the statement block of Example 4.22 making use of the FOR statement. The field index is incremented by the increment value of the control variable.

The control variable Index is initialised to 1. After every loop, Index is incremented by 1 (BY 1). It is possible to omit this "BY 1" part because 1 is the default increment value. To decrement the control variable, the value after BY must be negative.

A FOR statement consists of the following steps:

1) Initialise the control variable.
2) Check the end condition and stop execution of the loop if the control variable is outside the defined range.
3) Execute the statement block.
4) Increment the control variable (default is 1).
5) Continue with step 2.

Additional conditions for the FOR statement:

1) The control variable, the initial, end and increment value must have the same integer data type (SINT, INT or DINT).
2) It is forbidden to change the value of the control variable and all variables defining the start and end values within the statement block.
3) The end condition is checked before each loop.

It is allowed but dangerous to change the increment value inside the loop. IEC 61131-3 says nothing about when the compiler determines the new increment value (at the beginning or end of the loop). Therefore different programming systems could produce different results.

The value of the control variable on completion of the FOR statement is implementation-dependent. It is either the value assigned during the last loop or an incremented or decremented value.

EXIT statement.

Every iteration statement (REPEAT, WHILE and FOR) can be left early with the EXIT statement. In the case of nested statements, the inner iteration statement is left with EXIT and the following command of the next outer loop is executed.

```
WHILE (...)
    ...
    FOR ... TO ... BY ... DO
        ...
        IF ... THEN EXIT; ──────────────┐
                                         │  (* Continue after FOR Loop *)
        END_IF;                          │
        ...                              │
    END_FOR;                             │
    ... ◄────────────────────────────────┘
END_WHILE;
```

Example 4.24. Example of interrupting an iteration statement within a statement block using the EXIT statement.

4.2.8 Example: Stereo cassette recorder

The following example describes the control program of a *stereo cassette recorder* in ST. Section 4.3.5 shows a graphical solution in FBD and an explanation of the algorithm.

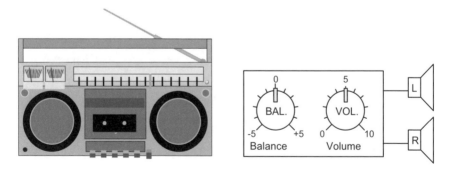

Figure 4.14. Example of a stereo cassette recorder with control elements for volume and balance.

The control program has to consider:

1) Adjustment of the two speakers depending on the current balance control setting (say an integer value between –5 and +5) and the volume control setting (say an integer value between 0 and +10). The amplifier output needs a REAL data type.
2) Volume control. If the volume exceeds a pre-defined constant value for some time, a warning LED must be turned on. Additionally a warning message is sent to the calling program.
3) There are two models of this recorder with different limits of loudness

```
FUNCTION_BLOCK Volume
  VAR_INPUT
    BalControl:     SINT (-5..5);      (* balance control with integer range -5 to 5 *)
    VolControl:     SINT (0..10);      (* volume control with integer range 0 to 10 *)
    ModelType:      BOOL;              (* 2 model types; TRUE or FALSE *)
  END_VAR
  VAR_OUTPUT
    RightAmplif:    REAL;              (* control variable for the right amplifier *)
    LeftAmplif:     REAL;              (* control variable for the left amplifier *)
    LED:            BOOL;              (* warning LED on: 1; off: FALSE *)
  END_VAR
  VAR_IN_OUT
    Critical:       BOOL;              (* return value *)
  END_VAR
  VAR
    MaxValue:       REAL := 26.0;      (* max. amplifier input; active for a defined time: *)
                                       (* turn on the warning LED *)
    HeatTime:       TON;               (* standard FB (time delay) to control *)
                                       (* the overdrive time *)
    Overdrive:      BOOL;              (* overdrive status *)
  END_VAR
```

```
(* Control of the right amplifier
depending on the volume and balance control settings *)
RightAmplif :=   Norm (           LCtrlK := VolControl,
                                  BIK := BalControl+5,
                                  MType := ModelType);

(*Control of the left amplifier,
balance control knob behaves in reverse to that of the right amplifier*)
LeftAmplif :=   Norm  (           LCtrlK := VolControl,
                                  BIK := ABS(BalControl - 5),
                                  MType := ModelType);

(* Overdrive ? *)
IF MAX(LeftAmplif, RightAmplif) >= MaxValue
THEN
        Overdrive := TRUE;
ELSE
        Overdrive := FALSE;
END_IF;

(* Overdrive for more than 2 seconds? *)
HeatTime (IN := Overdrive, PT := T#2s);

LED := HeatTime.Q;
IF HeatTime.Q = TRUE THEN
        Critical := 1;
   END_IF;
END_FUNCTION_BLOCK


FUNCTION Norm: REAL
  VAR_INPUT
      BIK:             SINT;    (* scaled balance control *)
      LCtrlK:          SINT;    (* volume control *)
      MType:           BOOL;    (* 2 types; described by TRUE or FALSE *)
  END_VAR
  TYPE
        CalType :REAL := 5.0; (* data type with special initial value *)
  END_TYPE
  VAR
    Calib:            CalType; (*Scaling value for amplifier output; initialised with 5.0 *)
  END_VAR

(* Evaluate real numbers for the amplifiers depending on model and control knob settings *)
   Norm :=    SINT_TO_REAL(BIK) +                  (* take the balance value *)
              Calib +                              (* add scaling value *)
              SEL(G := MType, IN0 := 4.0, IN1 := 6.0) +    (* model-specific scaling value *)
              SINT_TO_REAL(LCtrlK);                (* add the volume value *)

END_FUNCTION
```

Example 4.25. ST program with two POUs to solve the control problem of Figure 4.14.

4.3 Function Block Diagram FBD

The language *Function Block Diagram* (*FBD*) comes originally from the field of signal processing, where integer and/or floating point values are important. In the meantime, it has become established as a universally usable language in the field of industrial controllers.

The following section also applies to Ladder Diagram (*LD*). It describes the basic ideas and common elements of these two languages.

4.3.1 Networks, graphical elements and connections of LD and FBD

The representation of a POU using the graphical languages FBD or LD includes parts like those in the textual languages:

1) A leading and ending part of the POU,
2) The declaration part,
3) The code part.

The **declaration part** can be either graphical or textual. Most programming systems support, at least, textual declaration.

The **code part** is divided into **networks**. Networks are helpful to structure the control flow of a POU.

A *network* consists of:

1) Network label,
2) Network comment,
3) Network graphic.

Network label.
Every network (or group of networks) that is entered by a jump from another network has a prefix in the form of a user-defined alphanumeric identifier or an unsigned decimal integer. This is called the network label.

Programming systems usually number the networks consecutively. The consecutive numbering of **all** networks is updated automatically when a new network is inserted. This makes networks faster to find (e.g. for "GoTo network" or positioning at a reported error line) and corresponds to the line numbers in textual languages.

The standard only defines the network jump label. This label is an identifier local to one POU.

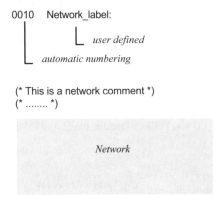

0010   Network_label:

       user defined

    automatic numbering

(* This is a network comment *)
(* .... *)

Network

Example 4.26. Allocation of network labels for the identification and marking of jump destinations

Network comment.
Between the network label and the network graphic, most programming systems allow the input of a comment delimited by (*...*) as in the textual languages.

IEC 61131-3 itself does not define a comment (and therefore does not prescribe the position or the format) for the graphical languages.

Network graphic.
The network graphic consists of *graphical objects* that are subdivided into *graphical (single) elements* as well as connections and connecting lines.

Information (data) flows via connections into graphical elements for the specified processing. The result is stored in the output parameters ready for sending to the next element.

To represent this flow of information, the elements are linked by lines which can also cross one another. The standard defines line crossings **without connections** (the lines are independent of each other and have no mutual influence) and crossings where lines are **joined** or **split** (the information flows together or is passed to more than one destination):

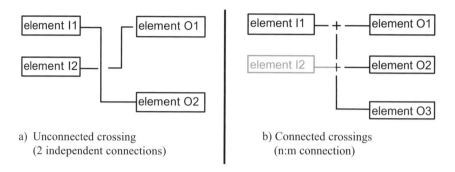

a) Unconnected crossing   b) Connected crossings
   (2 independent connections)   (n:m connection)

Figure 4.15. Crossing of connecting lines in LD and FBD. A junction (right: el I1 - el I2) is not possible in FBD.

In Figure 4.15, the exchange of information in case a) from I1 to O2 and I2 to O1 is mutually independent. In case b) all elements O1, O2 and O3 receive the same information from I1 and I2, as is shown by the "+" sign, or direct connection. The output lines of I1 and I2 are combined and the value is distributed to the input lines of O1, O2 and O3. A junction of lines like I1/I2 in case b) is only permissible in Ladder Diagram (dashed line), where it represents a "wired OR".

Some programming systems restrict networks to the maximum size of the screen or a print page; no horizontal scrolling is possible. For these systems, the standard helps with a construct called a **connector**. Please do not confuse this with a network label.

Connectors are not control or data flow elements, but the equivalent of a "new line" in a graphic. To construct long networks, it is possible to draw a named line at the right-hand edge of the screen. This named line appears again as an incoming line at the left-hand edge of the screen to show the continuation of the network. A named line like this is called a connector.

In Figure 4.16 the connector conn1 is used to divide the network network_x into two parts.

Connector names are POU-local identifiers. It is forbidden to use a connector name again as a network label or variable name (in the same POU).

Connectors can be defined by the user ("fixed line feed") or automatically by the programming system when displayed on the screen or print page. Some systems do not need these connectors because of their unlimited network depth.

All network elements connected directly with each other or indirectly via connectors belong to the same network.

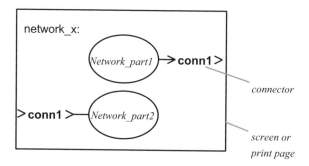

Figure 4.16. Connector to continue a network on a "new line" at the left-hand side of the screen (or paper) in LD and FBD

4.3.2 Network architecture in FBD

The graphical elements of an *FBD network* include rectangular boxes and control flow statements connected by horizontal and vertical lines. Unconnected inputs of the boxes can have variables or constants attached to them. The inputs/outputs can also remain open (free).

0001   StartNetwork:
(* network comment *)

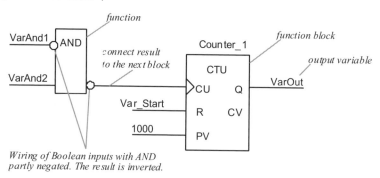

Example 4.27. Elements of an FBD network

Example 4.27 performs the AND function on the input variables VarAnd1 (inverted) and VarAnd2, and connects the inverted result with the counting input CU of Counter_1 (instance of FB type CTU). The output parameter Counter_1.Q is stored in a variable named VarOut.

Negated Boolean inputs/outputs of functions and FBs are marked with a circle "o".

Edge-triggered inputs of an FB are shown by ">" (rising edge) or "<" (falling edge) beside the formal parameter name.

IEC 61131-3 does not define whether the outgoing line of a function should appear at the top or bottom of the box (see AND of Example 4.27). It is also implementation-dependent whether the variable names are always at the left-hand edge of the page or next to the box or whether they should be written above or beside the connecting line. Example 4.28 shows an alternative layout to Example 4.27.

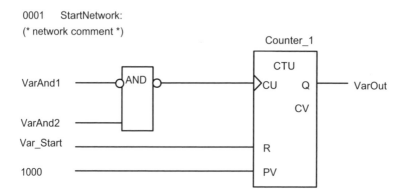

Example 4.28. Alternative FBD layout of Example 4.27

The graphical combination of functions and connections enables logical and arithmetic relationships to be represented clearly, see subsequent examples:

Example 4.29. A Boolean and an arithmetic expression represented using the elements of FBD

4.3.3 Graphical objects in FBD

FBD has the following *graphical objects*:

1) *Connections*,
2) *Graphical elements* for execution control (like JUMP),
3) *Graphical elements* to call a function or function block (standard or user-defined),
4) *Connectors*, see also Section 4.3.1.

FBD does not have any other graphical elements such as coils (see Ladder Diagram, Section 4.4.3).

The graphical objects of FBD are explained in the following sections and combined in Example 4.33 (FB Volume).

Connections.
FBD uses horizontal and vertical *connecting lines*. It is possible to split a line into several connections.

It is forbidden to connect more than one output with one or more inputs (junction). This would produce an inconsistency because it would be unclear which output should be routed to the inputs. This is only possible in Ladder Diagram ("wired OR").

Variables of elementary and derived data types can be passed on. It is important that the output parameter has the same data type as the connected input parameter(s).

Graphical object	Name	Explanation	Ex. [a]
	Horizontal connection	The horizontal connection copies the value written on the left-hand side to the right-hand side.	all
	Vertical connection with horizontal connections.	The horizontal connection copies the value written on the left-hand side to all places on the right-hand side (splitting).	[0004]
	forbidden connection ("wired OR")		

a   The number in brackets [...] is the number of the network in Example 4.33 using this feature

Table 4.8. Connections in FBD

Execution control (jumps).

In order to influence program execution, there are calls for leaving the POU (returns) and for changing the sequence of processing networks (jumps to network labels).

Graphical object	Name	Explanation	Ex. [a]
1─⟨RETURN⟩ (equivalent representation)	(Unconditional) return	The standard does not define any specific graphical element. The user can program an unconditional return to the calling POU by using a conditional return with input parameter TRUE (equivalent "1"). Control is also returned to the calling POU when the end of a POU is reached.	[0005]

nwp^b─⟨RETURN⟩	Conditional return	Left link of the RETURN element
		  == 1: Stop POU and return to the calling POU,
		  else: Ignore.
			[0005]

1 ─≫ Lab	(Unconditional) jump	Jump to the network labelled with the name "lab"	

nwp^b─≫ lab	Conditional jump	Left link of the RETURN element
		  == 1: Jump to the network labelled with the name "lab",
		  else: Ignore.
			[0004]

a   The number in brackets [...] is the number of the network in Example 4.33 using this feature
b   nwp defines a part of a network that supplies a Boolean output

Table 4.9. Graphical elements for execution control in FBD

Call of functions and function blocks.

The graphical representations for calling POUs of types FB and function are similar. FB and functions can have several input and output parameters. The formal parameters are written inside the box; extensible/overloadable functions do not have any formal parameter names, see Example 4.27. An *actual parameter* is a variable name or constant written next to (outside) its formal parameter. The second method of passing information is to connect the input parameter with the output of another box.

In FBD, it is **not** mandatory for FBs to have at least one Boolean input and output parameter. This requirement exists in Ladder Diagram.

The use of *EN/ENO* input and output parameters for function calls is possible but not required by IEC 61131-3; see Section 2.7.1. Many actual programming systems do not include this feature.

The *return value* of a function is determined by a value assignment to the function name (see output of the addition in network 0001 of function Norm, Example 4.33). If several assignments to the function name are made, the last value assigned before leaving the function is valid.

The return value in the function box is passed to the right-hand side and connected without a formal parameter name. The other output parameters are represented with formal parameter names (inside the box) for better readability.

In case of a VAR_IN_OUT declaration, a horizontal line connects the formal input and output parameter.

0001   Network_example:

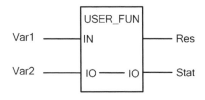

Example 4.30. Function call in FBD with a VAR_IN_OUT variable declared as IO and a return value stored in the variable Res.

4.3.4 Programming methods in FBD

Network evaluation.
A POU is evaluated network by network, from top to bottom. If it is necessary to change this computation sequence, FBD offers the possibility of jumps and returns (conditional and unconditional).

There are also deviating implementations where the evaluation of all networks is processed at the same time or according to execution sequence numbers. IEC 61131-3 leaves the sequence of network evaluation to the discretion of the programming system manufacturer; the chosen method must be documented in the manufacturer's manual.

All systems must adhere to the following provision: a single network (in FBD or LD) is evaluated by PLC according to the following rules:

1) Evaluate all inputs of a network element before executing the element.
2) The evaluation of a network element is not completed until all outputs of this element are evaluated.
3) The evaluation of a network is not completed until all outputs of all elements of the network are evaluated.

For factors affecting *cross-compilation*, see Section 7.3.1.

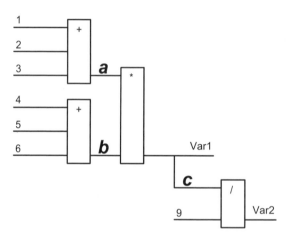

Example 4.31. Value calculation in an FBD network. a, b and c are intermediate values which are explained in the text below.

The sequence of evaluation in Example 4.31 follows the rules above: the input of the division (connecting line c) must be stable before the calculation of Var2 can be performed. To do this, the inputs of the multiplication function must first be evaluated. The sequence is: calculate the output of the first and second additions (constants are stable). Use the two outputs 6 (connection a) and 15 (connection b) as valid input parameters to the multiplication. The result, 90, is now stable. Store this value in the variable Var1. The division is now ready for evaluation. Use 90 (c) and the constant 9 to compute the value, 10, and store the result in the variable Var2. The sequence of storing Var1 and Var2 is not defined by IEC 61131-3.

Lines can connect all valid data types like Integer or String as long as both ends have the same data type. This is similar to type checking in IL using the Current Result, Section 4.1.2.

Feedback variable.

In FBD, the value of an output parameter can flow back to an input parameter of the same network. Such lines are called *feedback paths*; the associated variables are called *feedback variables*. (It is also possible to use the output parameters of a function block as feedback variables.) The first time such an input variable is evaluated, it has the default initial value for that data type. After the first cycle, its value is the corresponding output variable value from the previous cycle.

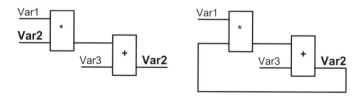

Example 4.32. FBD network with feedback variable Var2. There are two equivalent representations. On the left: implicit connection. On the right: explicit feedback by a connecting line.

There are some restrictions on using feedback variables with FB calls. These restrictions also apply to Ladder Diagram and are explained in Section 4.4.4.

4.3.5 Example: Stereo cassette recorder

The following example describes the *stereo cassette recorder* example of Section 4.2.8 (which contains both the problem description and the equivalent ST program).

A discussion of the program functionality follows the graphical representation.

```
FUNCTION_BLOCK Volume
  VAR_INPUT
     BalControl:    SINT (-5..5);    (* balance control with integer range -5 to 5 *)
     VolControl:    SINT (0..10);    (* volume control with integer range 0 to 10 *)
     ModelType:     BOOL;            (* 2 model types; TRUE or FALSE *)
  END_VAR
  VAR_OUTPUT
     RightAmplif:   REAL;            (* control variable for the right amplifier *)
     LeftAmplif:    REAL;            (* control variable for the left amplifier *)
     LED:           BOOL;            (* warning LED on: 1; off: FALSE *)
  END_VAR
  VAR_IN_OUT
     Critical:      BOOL;            (* return value *)
  END_VAR
  VAR
```

```
     MaxValue:      REAL := 26.0;        (* max. amplifier input; active for a defined time: *)
                                         (* Turn on the warning LED *)
     HeatTime:      TON;                 (* standard FB (time delay) to control *)
                                         (* the overdrive time *)
     Overdrive:     BOOL;                (* Overdrive status*)
END_VAR
```

0001:
(* Control of the right amplifier. This *)
(* depends on the volume and balance control settings *)

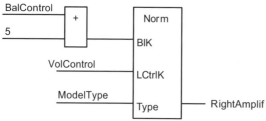

0002:
(*Control of the left amplifier, *)
(* balance control knob behaves in reverse to that of the right amplifier *)

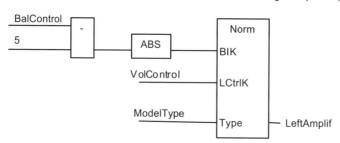

0003:
(* Overdrive ? *)

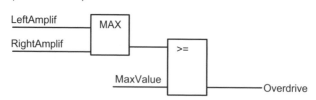

0004:
(* Overdrive for more than 2 seconds? *)

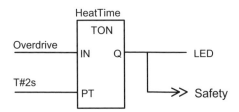

0005:
(*exit current POU *)

1 ─⟨RETURN⟩

0006 Safety:
(* Alternatively to the direct line: box with assignment symbol ":=" *)

1 ── Critical

END_FUNCTION_BLOCK

FUNCTION **Norm**: REAL
    VAR_INPUT
        BlK: SINT;   (* scaled balance control *)
        LCtrlK: SINT;   (* volume control *)
        MType: BOOL;   (* 2 types; described by TRUE or FALSE *)
    END_VAR
    TYPE
        CalType :   REAL := 5.0; (* data type with special initial value *)
    END_TYPE
    VAR
        Calib: CalType; (*Scaling value for amplifier output; initialised with 5.0 *)
    END_VAR

    0001:
    (* Evaluate real numbers for the amplif. depending on model and control knob settings *)

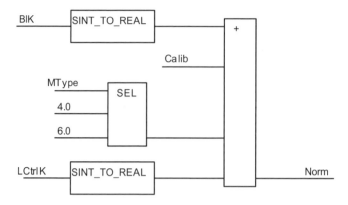

END_FUNCTION

Example 4.33. FBD program with two POUs to control the stereo cassette recorder. See also the equivalent program written in ST, Section 4.2.8.

Comments on the networks of Example 4.25 and Example 4.33

Network-number	Comment

FB Volume:

0001	The network calls the function Norm, which receives as actual parameters the sum of BalControl and 5 (scaling from -5..5 to 0..10), the value of the volume control and the Boolean identification of the ModelType. If there were more than two model types, an enumerated data type would be a better choice than Boolean. A floating-point number is the result (right / left amplifier).
0002	See network 0001. Because the balance control knob has exactly the reverse effect as for the right amplifier, the settings are scaled from -5..5 to 10..0.
0003	Check to see whether one of the two amplifier values exceeds the maximum value. Overdrive receives the Boolean value of the check.

0004   If Overdrive has the value TRUE, the instance HeatTime of the
                  standard FB TON is started or continues running. HeatTime
                  provides TRUE at the output Q after 2 seconds. It is reset as soon
                  as FALSE is stable at the input parameter IN.
                      This network turns the LED on and off.
                      If the output Q is TRUE, the program jumps to Safety (via the
                  conditional jump).

0005   Return to the calling program.

0006   Response of TRUE to the calling program via the IN_OUT
                  variable Critical indicates that the volume is in a critical region.

Function Norm:
0001   The "+" adds the following: 1) balance values converted from
                  Short Integer to Real, 2) a calibrating value, 3) one of two
                  alternative values, depending on the input parameter MType, and
                  4) the volume.
                  The result is a floating-point numeric function value.

4.4 Ladder Diagram LD

The language *Ladder Diagram* (LD) comes from the field of electromechanical
relay systems and describes the power flow through the network of a POU from
left to right. This programming language is primarily designed for processing
Boolean signals (1 ≡ TRUE or 0 ≡ FALSE).

4.4.1 Networks, graphical elements and connections (LD)

Like FBD, LD also uses networks. They are always processed in sequence from
top to bottom, unless specified otherwise by the user. The description of the basic
elements and connections can be read in Section 4.3.1.

4.4.2 Network architecture in LD

An *LD network* is bounded by so-called power rails on the left and on the right.
From the left rail, "powered" by logic state 1 ("power flow") power reaches all
connected elements. Depending on their logic state, elements either allow the
power to pass on to the following connected elements or they interrupt the flow.

0001   StartNetwork:
(* Network comment *)

Var1   Var2   VarOut

├──┤├──────┤├──────( )──┤

Code part with left and right power rail

Example 4.34. Graphical elements of an LD network

4.4.3 Graphical objects in LD

In LD, the evaluation of a network result depends on the graphical layout of the network (including variable names) and its connections. Elements are either connected in series or in parallel.

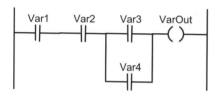

Example 4.35. The elements Var1, Var2 are connected in series (AND) with the parallel construct of Var3, Var4 (OR function). Var1 to Var4 are called "contacts" (variables are read), VarOut is called a "coil" (variable is written).

The network of Example 4.35 evaluates the logical expression:
VarOut := Var1 ∧ Var2 ∧ (Var3 ∨ Var4).

An LD network consists of the following graphical objects:

1) Connections,
2) Contacts and coils,
3) Graphical elements to control the execution sequence (jumps),
4) Graphical elements to call a function or FB (standard or user POU),
5) Connectors, see Section 4.3.1 .

These graphical elements are explained in the following sections:

Connections.
Like FBD, LD has horizontal and vertical lines to connect elements, and also includes crossing points.

Graphical object	Name	Explanation	Ex. [a]
———	Horizontal connection	The horizontal connection copies the value written on the left-hand side (FALSE or TRUE) to the right-hand side.	
	Vertical connection including horizontal lines	A vertical connection evaluates all incoming lines (from left) with OR and copies this value to all connections on the right-hand side.	[0005]

a   The number in brackets [...] is the number of the network in Example 4.44 using this feature

Table 4.10. Connections in Ladder Diagram

Contacts and coils.

A contact performs a logic operation on the value of an incoming line and the value of the associated variable. The kind of logic operation depends on the type of the contact. The calculated value flows to the right connecting line.

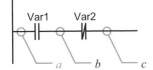

Var1	Var2	Not Var2	value of b	value of c
0	0	1	0	0
1	0	1	1	1
0	1	0	0	0
1	1	0	1	0

Example 4.36. Open contact and closed contact in AND operations: Connection a has the value TRUE (left power rail always TRUE). b is the result of "Var1 AND a (TRUE)", c is the result of "b AND NOT Var2 ".

Var1	Var2	Var3	value of a	value of b
0	0	0	0	0
1	0	0	0	0
0	1	0	0	0
1	1	0	1	0
0	0	1	0	0
1	0	1	1	1
0	1	1	0	0
1	1	1	1	1

Example 4.37. Three open contacts in an AND/OR operation: Connection a has the result of "Var1 AND (Var2 OR Var3)"; b has the value "Var1 AND Var3".

Contacts (open and closed) control the passing on of values in a network. The values passed depend on the value of the contact variables as well as their position in the network (the values of the contacts themselves are not changed!). Results are assigned to output variables by means of coils (values are assigned to the coil variables).

The name of the variable belonging to the contact is written above the graphical element. All variables have the data type Boolean (TRUE / FALSE). Consequently, the corresponding connections transmit only Boolean values.

There are two kinds of contacts: Normally open and Normally closed. An *Normally open contact* is like an electromechanical contact: when it is not activated (FALSE), it is open, and when it is activated (TRUE) it closes and power can flow. A *Normally closed contact* is the exact opposite: power flows when it is not activated (FALSE). It opens and interrupts the power flow when it is activated (TRUE).

Graphical object	Name	Explanation	Ex. [a]
VarName —||—	Normally open contact	RightLink := LeftLink AND VarName, (Copy the status of the left link to the right link if the variable evaluates to TRUE, else FALSE).	[0007], [0009]
VarName —|/|—	Normally closed contact	RightLink := LeftLink AND NOT VarName, (Copy the status of the left link to the right link if the variable evaluates to FALSE, else FALSE).	[0007], [0008], [0010]

VarName
—|P|—
	Positive *transition-sensing contact*	RightLink
		:= TRUE, if
		  * the variable had the value FALSE during the last evaluation (cycle) **and**
		  * the variable now has the value TRUE (current cycle) **and**
		  * LeftLink has the status TRUE
		:= FALSE otherwise.
		(Copy the value of the left link to the right link when the variable has changed from FALSE → TRUE.
		For all other combinations: copy FALSE to the right link).
			[0001],
			[0005],
			[0006]

VarName
—|N|—
	Negative transition-sensing contact	RightLink
		:= TRUE, if
		  * the variable had the value TRUE during the last evaluation (cycle) **and**
		  * the variable now has the value FALSE (current cycle) **and**
		  * LeftLink has the status TRUE
		:= FALSE otherwise.
		(Copy the value of the left link to the right link when the variable has changed from TRUE → FALSE.
		For all other combinations: copy FALSE to the right link).	

a   The number in brackets [...] is the number of the network in Example 4.44 using this feature

Table 4.11. Ladder Diagram: contacts

Coils are elements where values are assigned to variables:

Graphical object	Name	Explanation	Ex. [a]
VarName —( )—	Coil	Variable [b] := LeftLink. (Copy the value of the left link to the variable).	[0007]
VarName —(/)—	Negated coil	Variable [b] := NOT LeftLink. (Copy the negated value (change FALSE <-> TRUE) of the left link to the variable).	

VarName
—(S)—
	Set (latch) coil	Variable [b]
		    := TRUE, if LeftLink has the state TRUE,
		    else: do not change the variable.
			[0010]

VarName
—(R)—
	Reset (unlatch) coil	Variable [b]
		    := FALSE, if LeftLink has the state TRUE,
		    else: do not change the variable.
			[0005]

VarName
—(P)—
	Positive transition-sensing coil	Variable [b]
		    := TRUE, if
		        * LeftLink had the status FALSE during the last evaluation (cycle)
		        and
		        * LeftLink now has the status TRUE (current cycle)
		    else: do not change the variable.
		(Copy the value of the left link to the variable when the left link has changed from FALSE → TRUE.
		Otherwise: do not change the variable.)	

a   The number in brackets [...] is the number of the network in Example 4.44 using this feature
b   Always copy the left link to the right link

Table 4.12. Ladder Diagram: coils (continued on next page)

VarName —(N)—	Negative transition-sensing coil	Variable [b] := TRUE, if * LeftLink had the status TRUE during the last evaluation (cycle) **and** * LeftLink now has the state FALSE (current cycle) else: do not change the variable. (Write TRUE to the variable if the left link has changed from TRUE → FALSE. Otherwise: do not change the variable.)	[0009]

a   The number in brackets [...] is the number of the network in Example 4.44 using this feature
b   Always copy the left link to the right link

Table 4.12. (Continued)

Execution control.

To define the sequence of execution of POUs and their networks, LD offers two types of *execution control*: one to leave a POU and the second to jump to a specified network.

Graphical object	Name	Explanation	Ex. [a]
⊢<RETURN>	(Unconditional) return	Leave the POU and return to the calling POU. Execution control returns automatically to the calling program when the end of a POU is reached.	[0011]
nwp[b]<RETURN>	Conditional return	LeftLink of the RETURN element == 1: Stop the POU and return to the calling POU, else: Ignore.	

a   The number in brackets [...] is the number of the network in Example 4.44 using this feature
b   nwp represents a part of a network which supplies a Boolean value

Table 4.13. LD: Graphical objects for execution control (continued on next page)

>> Label	(Unconditional) jump	Jump to the network labelled with the name "Label"	[0002], [0004]
nwp[b] >> Label	Conditional jump	LeftLink of the jump element == 1: Jump to the network labelled with the name "Label", else: Ignore.	[0001], [0008]

a   The number in brackets [...] is the number of the network in Example 4.44 using this feature
b   nwp represents a part of a network which supplies a Boolean value

Table 4.13. (Continued)

Call of functions and function blocks.
LD supports calls to POUs of type Function or FB.

The called POU is represented by a rectangular box. FBs and functions can have several input and output parameters. Functions can have one or more return parameters. The names of the formal input and output parameters appear inside the box; extensible/overloadable functions do not have any formal parameter names. The appropriate *actual parameters* (variables, constants) are written on the connecting line outside the box adjacent to the *formal parameter* name on the inside. It is also possible to connect inputs directly with input or output parameters of other boxes (equivalent to FBD).

The negation of inputs/ outputs and the marking of edge-triggered inputs are equivalent to FBD; see Section 4.3.2.

The parameters of **function blocks** can be of any data type, but at least one input and one output parameter must be of type Boolean and these must have a direct or indirect connection to the left and right power rails. Standard FBs have the output Q to meet this requirement.

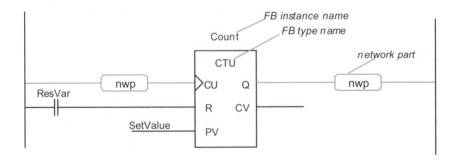

Example 4.38. Call of the standard FB CTU in an LD network

The variable SetValue is assigned to the parameter PV. The value of ResVar is copied to the input R. Output CV is not connected. The Boolean result of the left nwp controls the transition-sensing input CU of Count. The output parameter Q is the start value of the right nwp (a Boolean part of an LD network).

The left nwp in Example 4.38 is not mandatory because there is an existing connection in this example to the left power rail (input R via ResVar). The right nwp is necessary because CV is an integer parameter and cannot be connected to the right power rail.

Functions must have special input and output parameters called EN / ENO controlling the execution of the function. If EN evaluates to FALSE, the function is not executed and the outgoing parameter ENO is also set to FALSE. It is possible to use ENO as an error parameter (see Section 2.7.1). EN and ENO must have a direct or indirect connection (via a network part) to the left and right power rails. The state of the other output parameters for ENO=FALSE is manufacturer-specific and must be documented.

It is implementation-dependent whether the other parameters of a function (e.g. IN0 in Example 4.39) are pre-set by a variable or a network part.

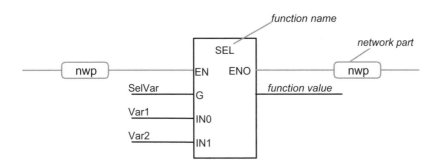

Example 4.39. Call of the standard function SEL in an LD network

The *function value* of a user-defined function is determined by the assignment to a variable (coil) with the name of the function (see other IEC 61131-3 languages).

4.4.4 Programming methods in LD

Network evaluation.
The execution sequence of LD *networks* is identical to FBD networks, i.e. from top to bottom. To change this execution sequence the user can program jumps.

In a network, IEC 61131-3 allows further processing of a value after (intermediate) assignment, as shown in Example 4.41. However, most programming systems use the following pattern:

- Left-hand part. A logical construct with variables, constants and functions from which a value is calculated (computing),
- Right-hand part for storing the result (saving).

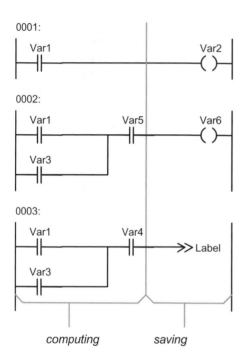

Example 4.40. Structure of a LD network

0000:

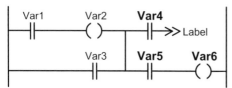

Example 4.41. A permitted but unusual network with the functionality of Example 4.40 using a scratch variable named Var2 in this example

The network of Example 4.41 is more compressed than the three networks of Example 4.40, but Example 4.40 gives a better overview of the logical functionality.

The Ladder Diagram language is an easy way to program Boolean expressions. If it is necessary to evaluate other data types, such as Integer or Real, it is better to use a different language. Some programming systems can incorporate the layout of such non-Boolean networks in the form of an FDB network; the rest remains in Ladder Diagram.

Feedback variable.
If a variable is used for storing a value in one cycle and read as an input for the same network in the next cycle, it is called a *feedback variable*.

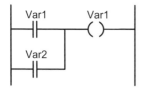

Example 4.42. Feedback variable (Var1) in LD

Unlike FBD, only implicit feedback connections (by repeating the variable name) are possible in LD. Therefore, no explicit links from right to left are allowed (compare this with the FBD example in Example 4.32).

If FBs and feedback variables are part of the same network, this can result in different behaviour for different PLC systems:

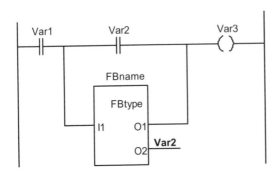

Example 4.43. Network including an FB and a feedback variable (Var2). The result is system-dependent; see the text below.

In Example 4.43, it is unclear whether the contact Var2 is assigned the value of the parameter FBname.O2 from the last or current cycle. A comparable situation occurs if, in Example 4.41, variable Var4 is replaced by Var6. It is ambiguous whether the contact Var6 is assigned the current value of the coil Var6 or that of the last network evaluation.

IEC 61131-3 does not discuss this ambivalence in detail. In order to exclude such implementation dependencies, an FB should be regarded as a network of its own. If there is a need to use parts of networks to predefine input parameters or to use output parameter values in the same network, the same variable should not be assigned to a contact and a coil in the same network. In short, do not write networks like Example 4.41!

The sequence problem described in Example 4.43 also applies to global data. If Var2 in FBname is of type EXTERNAL and there is a write instruction to Var2 within the body of Fbname, the value of the contact Var2 is unclear (Is it the value of the assignment inside the FB or the value of the output parameter?). If an FB influences its calling environment using EXTERNAL variables in this manner, the FB produces *side effects*. This should be avoided.

If a network does not need information from the last cycle, it is better to use function calls instead of FB calls. If there is no way to avoid FBs in a network, check carefully the use of EXTERNAL variables within the FB body for this problem.

4.4.5 Example in Ladder Diagram: Mountain railway

The following example is a representation of the *mountain railway* example in Section 4.1.5 (IL) in LD, which also describes the requirements of this control task.

A detailed description of the networks follows the graphical representation, see below.

```
FUNCTION_BLOCK MRControl
VAR_INPUT
    MRStart:        BOOL;          (* Button to start the railway *)
    MREnd:          BOOL;          (* Switch for initiation of the end of operation *)
    S1, S2, S3:     BOOL;          (* Sensors in every station *)
    DoorOpenSignal:BOOL;           (* Switch to open the door. 1: open; 0: close! *)
END_VAR

VAR_IN_OUT
    StartStop:      BOOL;          (* Cabin moving: 1; not moving: 0 *)
END_VAR

VAR_OUTPUT
    OpenDoor:       BOOL;          (* Motor to open the door *)
    CloseDoor:      BOOL;          (* Motor to close the door *)
END_VAR

VAR_OUTPUT RETAIN
    EndSignal:      BOOL;          (* Warning signal for Power Off *)
END_VAR

VAR
    StationStop:    CTU;           (* Standard FB (counter) for cabin stops *)
    DoorTime:       TON;           (* Standard FB (delayed) start of cabin *)
    DirSwitch:      SR;            (* Standard FB (flip-flop) for change of direction *)
END_VAR

VAR RETAIN
    Direction:      BOOL;          (* Current direction up or down *)
END_VAR
```

0001:

(* System running for the first time after power on? Yes: Reset the end signal *)
(* activated by the last shutdown *)

```
       MRStart                                    EndSignal
      ──┤P├──────────────────────────────────┬────(R)────
                                              │
                                              └──≫  ResCount
```

0002:

0003   ResCount:
(* Reset the station counter *)

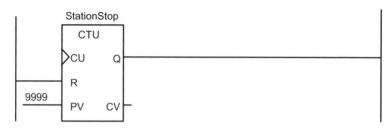

0004:

0005   Arrive:
(* Increase the counter StationStop when the cabin arrives at a station *)

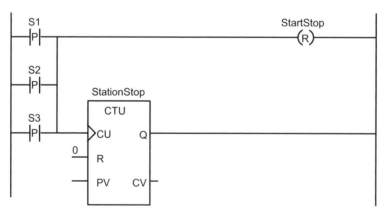

0006:
(* Change of direction? *)

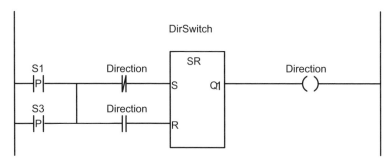

0007:
(* Condition to open cabin: cabin stops and door open switch is activated *)

0008:
(* End signal for railway and cabin in a station -> POU end *)

0009   CloseCabin:
(* Close door when door switch operated *)

0010:
(* Cabin start 10 seconds after activation of the door switch *)

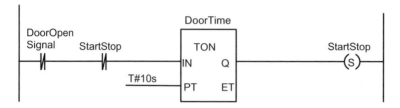

0011:
(* Return to the calling POU *)

END_FUNCTION_BLOCK

Example 4.44. LD program to control the mountain railway; see also Example 4.13.

Comments on the mountain railway networks.

Network-number	Comment
0001	A rising edge of variable MRStart causes TRUE to be sent to the right side. In this case, EndSignal is set to FALSE (retentive); then a jump is made to the network ResCount. If no rising edge is detected, the next network, 0002, is activated.
0002	Jump to network Arrive (network 0005).
0003	The input R is set to TRUE. This forces a reset of the instance StationStop of FB type CTU. The pre-set parameter PV is set to 9999. The output parameter Q is not computed. Continue with the next network 0004.
0004	Jump to network CloseCabin (network 0009).

0005   As soon as the cabin arrives at one of the three stations, the corresponding signal S1, S2 or S3 changes from FALSE to TRUE (rising edge). Via the OR function, CU is set to TRUE after such a transition (and only then). StationStop is then incremented by 1. The 0 at input parameter R is not necessary. In the case of a rising edge, the cabin stops, forced by StartStop ("reset coil" means: write FALSE to StartStop).

     If none of the signals S1-S3 has changed from FALSE→TRUE, FALSE is written to the parameter CU and StartStop remains unchanged.

0006   As soon as the cabin arrives in the valley or at the top station, one of the corresponding signals S1 or S3 changes from FALSE to TRUE. The OR operation evaluates in this situation (and only then) to TRUE. If the direction flag is TRUE, the RESET parameter is set to TRUE and therefore the direction flag is reset (via the flip-flop DirSwitch).

     Otherwise, the direction flag is changed via the SET parameter (from FALSE to TRUE).

0007   Logical AND operation on DoorOpen and the inverted value of StartStop.

DoorOpenSignal	StartStop	Not StartStop	OpenDoor
0	0	1	0
1	0	1	1
0	1	0	0
1	1	0	0

The door opens only when the door open button is set and the motor of the cabin has stopped.

0008   Logical AND operation on MREnd and the inverted value of StartStop.

MREnd	StartStop	Not StartStop	EndSignal
0	0	1	0
1	0	1	1
0	1	0	0
1	1	0	0

As long as the switch MREnd is TRUE and the cabin is standing at a station, the signal EndSignal is active (TRUE). The control returns to the calling POU.

     Otherwise the next network 0009 is activated.

0009   If DoorOpenSignal signals a TRUE→FALSE edge, CloseDoor is
                set to TRUE, otherwise there is no change.

0010   As soon as the door is closed (DoorOpenSignal evaluates to
                FALSE) and the cabin is in one of the three stations (StartStop
                evaluates to FALSE), IN is set to TRUE. Now the timer DoorTime
                starts. After 10 seconds, the output Q changes from FALSE to
                TRUE. If the door is still closed (DoorOpenSignal equal FALSE),
                the cabin starts (StartStop is TRUE). If the door has been opened
                in the meantime, DoorTime is reset.
                   If the timer is not active (output Q is FALSE) or the
                OpenDoorSignal evaluates to TRUE, StartStop is not changed.
                   StartStop is a feedback variable. The value of the associated
                closed contact is the initialisation value of the variable or the
                value assigned to the coil in a previous cycle.

0011   Return to the calling program. Programming a loop within this
                POU (jump to the first network of the POU) would not be the
                right solution because this POU needs updated parameters in
                every new cycle.

4.5 The American way of Ladder programming

The use of specific programming languages differs between countries and conti-
nents. German programmers prefer IL and ST; many of the PLC experts in France
structure their programs using SFC. In the US and Asia, a combination of SFC and
Ladder is the favourite programming platform.

During the last years these two languages have been used to develop many PLC
algorithms. The experience gained with these languages has influenced the
definition of the IEC 61131-3. However, the IEC 61131-3 languages and the
American languages cannot be integrated directly because the IEC 61131-3 includes some features like instantiation that are not part of the American languages.
So, we can find many similar features and some differences between the two; these
are discussed in the following sections.

Ladder and SFC are utilised nearly exclusively in the US. This exclusivity influences the supported features. The US Ladder is the base language not only for
Boolean expressions (as LD in IEC 61131-3)); it is also the means of programming
algorithms for any data type. This means that it needs additional features, not part
of IEC 61131-3, to support these other data types.

4.5.1 Network Layout

To program Integer or Real operations the EN/ENO (see Section 2.7.1) combination is used. The Boolean EN input connection prevents or enables execution of a module. The ENO output connection from a module controls the execution of the successor module in the same network; this largely corresponds to the definition in IEC 61131-3.

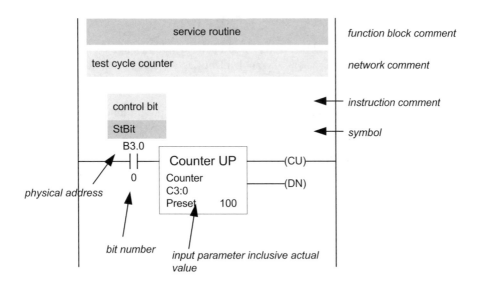

Example 4.45. Typical US- Ladder network inclusive function block controlled by an *EN / ENO* mechanism. Contact B3.0 is connected to the EN input and the coil CU stores the value of the ENO output. The use of EN / ENO is implicit i.e. their names do not appear on the module.

Between the two power rails, elements - similar to the IEC 61131-3 objects – are connected by sequential and parallel lines; see also Section 4.4.2. These elements are also called *open* and *closed contacts* (left side) or *coils* (right side) . Instead of one of these symbols, a POU (function or function block) can also be inserted.

(Formal and actual) input parameters are placed within the block. The parameter assignment is described by names (like Preset in Example 4.45) and a constant (100) or memory address (Counter -> C 3:0). Formal **and** actual parameters are both written inside the module frame. Simple coils on the right side receive the values of the output parameters. A pre- or post wiring of more than the first input or output parameter is not possible.

The American programming systems offer many types of comments. Titles and comments are allowed to describe:

- Module description.
- Network description.
- Symbols representing hardware addresses.
- Instructions.

4.5.2 Module addresses and memory areas

The instance address of a POU (referenced via its symbolic name) is not generated automatically. A number, not a symbol, identifies a module and this number is used to call the POU.

Variables are not administered automatically. The programmer identifies the memory areas by specially assembled names, see below. It is possible to assign symbolic names to these specified addresses to improve readability. Additionally, there are many pre-defined system variables; these have a well-defined behaviour such as system status, I/O areas or counter and timer parameters.

The name of a memory area (like I/O, intermediate result, parameter or standard FB) has a multi- stage hierarchy.

Data type→ file number → element → sub-element

examples: F5:11
             R3:5.DN
             I:016/05

Example 4.46. Hierarchical addressing of memory

The memory is divided in data type areas (basic types Integer, Real and Boolean; and additionally I/O, counters, timers and controls). The three examples of Example 4.46 address a floating-point memory ("F"), a control area ("R") and an input area ("I") respectively.

Every data type has a pre-defined number of data blocks (called "files") which include elements of the same data type. Such an element can be subdivided in additional sub-elements. The first example in Example 4.46 addresses the 11th element of the 5th file of the Real area. The second example specifies the 5th element of the 3rd control file; within this structure, the DONE Bit (DN) is addressed. The last example is an input image in rack number 01, group number 6, terminal number 05. It is obvious that there is a strong connection between naming convention and the hardware architecture.

Memory can be addressed directly (Example 4.46). Additionally it is possible to use *indirect addressing*. An indirect address specifies the memory where the address of the demanded cell is stored. This is described in the Example 4.47.

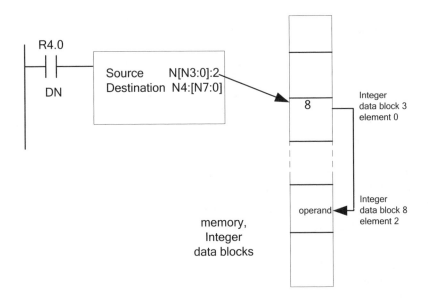

Example 4.47. Indirect addressing. The address references a memory where the actual address of the operand is stored. "[ ]" are used to indicate indirect addressing.

In Example 4.47 the integer storage cell, which is indirectly addressed via N3:0 , contains the value 8. This value 8 replaces N[3:0] to yield N8:2 as the desired source operand address. The destination is in the 4th integer data block; the associated element number is stored in data block 7, element 0.

Because the memory is specified directly by name, it is not necessary to declare variables as is demanded in IEC 61131-3. The numeric part of a name declares the physical address and the leading character specifies the data type (O: output; I: input; S: status; B: binary; T: timer; C: counter; R: control; N: integer; F: floating point (REAL)).

A variable is specified by its name determining the (elementary) data type. It is also possible to reference directly parameters of timers (T5.0) and counters (C3:0). The instances of these counters and timers run asynchronously to the user program (separate tasks). It is not necessary for the user to call these POUs in every cycle. They are scheduled by the system. This is different to the IEC 61131-3 standard function blocks. So it is possible to access the parameters of these standard blocks without calling the FBs by specifying the POU number and parameter name.

All programs and data blocks are identified by a number. This seems quite different to the instantiation concept of the IEC 61131-3.

It is possible to call a POU several times executing it with different data (just like the instantiation concept of the IEC 61131-3 where the same code is executed with different data declarations). This can be achieved by indirect addressing. Before calling the module, the required base address is written in some fixed address used by the POU. Each call of the POU has its own data space and the called POU addresses its data relative to this base address using an index (offset).

Let us have a look at such a construct: The address N[N5:0]:7 programmed in the called POU references an Integer value. The value can be found:

- by reading the number stored in address N5:0 and then
- by accessing the data block with this number. Element 7 of this data block contains the required data.

Each calling POU can change the contents of memory N5:0 to specify another data block. This new data block is then used by the called POU which references its data via the same fixed, indexed offsets.

4.6 Sequential Function Chart SFC

The IEC 61131-3 language SFC was defined to break down a complex program into smaller manageable units and to describe the control flow between these units. Using SFC, it is possible to design sequential and parallel processes.

The timing of the execution of these units depends on both static conditions (defined by the program) and dynamic conditions (behaviour of the I/Os). The units themselves are programmed in one of the other four IEC 61131-3 languages or in a further SFC structured description.

The methodology of SFC has been derived from well-known techniques like Petri-net or sequence (cascade) methodology.

The first widespread specification language in the automation market describing a system by several states and transition conditions was Grafcet (by Teleme-chanique, France). This language has been used as the basis for an international standard (IEC 848). With SFC, all these methodologies have been integrated into the world of IEC 61131-3.

Processes with a step by step state behaviour are especially suitable for *SFC structuring*. Take for example, a chemical process: liquids are added to a container until the sensors report that the container is full (step 1). A mixer is activated after closing the liquid valve (step 2). When the required mixing is finished, the mixer motor is stopped and the "empty container" process begins (step 3).

A second typical example of a step by step state behaviour is the washing machine. It works step for step beginning with a pre-wash cycle, main wash cycle, rinsing etc. Every step is controlled by a special algorithm. When the end of a step is reached (reported by a timer or sensor), a subsequent step with other serial and parallel control units is activated. The new set of instructions is executed (cyclically) until the end criterion of this step is signalled.

SFC is primarily a graphical language, although a textual description is also defined. This is mainly to allow the exchange of information between different programming systems or to simplify back-up of the POU. To develop the SFC structure of a POU, the graphical version is a better choice because it visualises the dependencies and interdependencies of the steps much more clearly.

It is possible, but not necessary, to structure a program or FB using SFC. It is not permissible to structure functions in SFC because an SFC POU needs static information, i.e. retained state information, which is not allowed in functions. If a POU is written in SFC, the **whole** POU must be structured in SFC.

The rules for *comments* within the SFC language are the same as for the other IEC 61131-3 languages. In the textual form, the parenthesis/asterisk combination (* **comment** *) can be written wherever blank spaces are allowed. There are no rules for the graphical form; comment form and type are implementation-specific.

4.6.1 Step / Transition combination

As in the other graphical languages, LD and FBD, the first level of structuring within the SFC language is the network. A POU has one or more networks. The elements of a network are called Steps and Transitions.

A *step* is either active or inactive. When a step is active, an associated set of instructions (called actions) is executed repeatedly until the step becomes inactive.

The decision to change the status of a step (from active to inactive or vice versa) is defined by a *transition*, which is the next element immediately below the step. A transition is programmed by a *transition condition*, which is a Boolean expression. When the expression becomes TRUE, the step is ready to be deactivated. One or more links below the transition describe which step or steps are activated when the current step becomes inactive.

Example 4.48 shows an *SFC network* consisting of steps (rectangular boxes), transitions (bars with identifiers) and links. The right-hand side of the diagram describes the operational sequence of the constructs.

When a POU structured in SFC is called, a special step called the initial step is activated. In the example, Start is the initial step. The set of instructions (not shown in the diagram) allocated to this step is executed. If the value of the Boolean variable StartReady changes from FALSE to TRUE due to computation of the step instructions or by an I/O change, Start is de-activated. At the same time, the next steps, Simultan1 and Simultan2, which are connected via the transition, are activated.

All instructions associated with these steps are now executed. After this, the value of the next transition is checked, i.e. the variable SimultanReady is evaluated. The Simultan steps are executed until the value of SimultanReady changes from FALSE to TRUE. When this occurs, the two Simultan steps are made inactive and the step Together becomes active. The value of the transition NewLoop now controls the state of the network and the step Together remains active until the Boolean expression of the NewLoop transition becomes TRUE.

With the triggering of a transition, the "*active*" *attribute* (often referred to as a *token*) is passed on from an active step to its successor(s). Consequently, this token "wanders" through the steps making up the network; it splits itself up in the case of a parallel branch and re-assembles itself again when the parallel branch terminates. The token may not "get lost", multiply or circulate in an uncontrolled manner through the network.

The graphical objects through which this token is passed are the subject of the next section.

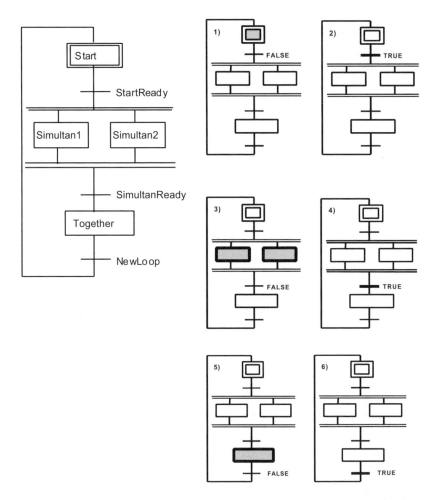

Example 4.48. Left: A simple SFC network. Right-hand side: Representation of the flow through this network in steps 1) to 6): Step (1) becomes inactive as soon as the subsequent transition evaluates to TRUE (2). This activates the following steps (3) and so on. When the flow reaches the end, the network begins again with the initial step (1).

4.6.2 Step - transition sequence

Steps and transitions must always alternate in a network. It is not permitted to link two elements of the same type. A transition can have several predecessor and successor steps. A simple combination of steps and transitions is called a "*Sequence*".

After executing a step it is possible either to select one of several subsequent sequences (alternative sequences) or to activate several sequences at the same time (simultaneous sequences). The rules governing these sequences are described in Table 4.14.

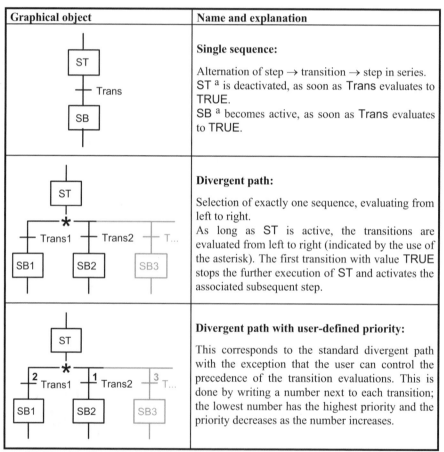

Graphical object	Name and explanation
	Single sequence: Alternation of step → transition → step in series. ST [a] is deactivated, as soon as Trans evaluates to TRUE. SB [a] becomes active, as soon as Trans evaluates to TRUE.
	Divergent path: Selection of exactly one sequence, evaluating from left to right. As long as ST is active, the transitions are evaluated from left to right (indicated by the use of the asterisk). The first transition with value TRUE stops the further execution of ST and activates the associated subsequent step.
	Divergent path with user-defined priority: This corresponds to the standard divergent path with the exception that the user can control the precedence of the transition evaluations. This is done by writing a number next to each transition; the lowest number has the highest priority and the priority decreases as the number increases.

a   ST: Step at the Top of the transition
SB: Step Below the transition

Table 4.14. Possible SFC sequences: Transitions and steps in alternation. Complex networks can be built up using combinations of these sequences. (Continued on next page)

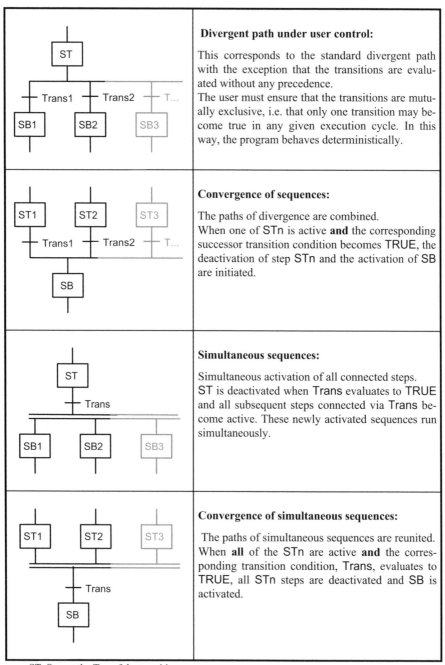

Divergent path under user control:

This corresponds to the standard divergent path with the exception that the transitions are evaluated without any precedence.

The user must ensure that the transitions are mutually exclusive, i.e. that only one transition may become true in any given execution cycle. In this way, the program behaves deterministically.

Convergence of sequences:

The paths of divergence are combined.
When one of STn is active **and** the corresponding successor transition condition becomes TRUE, the deactivation of step STn and the activation of SB are initiated.

Simultaneous sequences:

Simultaneous activation of all connected steps.
ST is deactivated when Trans evaluates to TRUE and all subsequent steps connected via Trans become active. These newly activated sequences run simultaneously.

Convergence of simultaneous sequences:

The paths of simultaneous sequences are reunited. When **all** of the STn are active **and** the corresponding transition condition, Trans, evaluates to TRUE, all STn steps are deactivated and SB is activated.

a   ST: Step at the Top of the transition
     SB: Step Below the transition

Table 4.14. (Continued on next page)

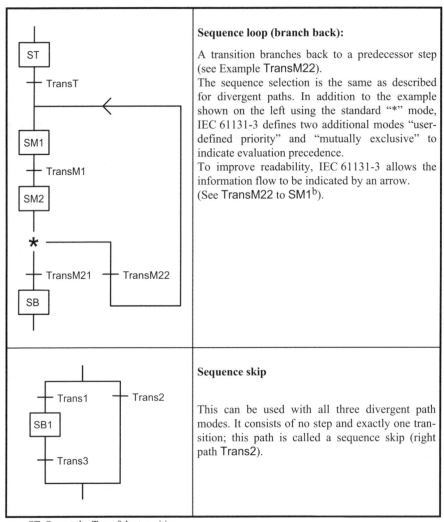

	Sequence loop (branch back):
	A transition branches back to a predecessor step (see Example TransM22). The sequence selection is the same as described for divergent paths. In addition to the example shown on the left using the standard "*" mode, IEC 61131-3 defines two additional modes "user-defined priority" and "mutually exclusive" to indicate evaluation precedence. To improve readability, IEC 61131-3 allows the information flow to be indicated by an arrow. (See TransM22 to SM1[b]).
	Sequence skip
	This can be used with all three divergent path modes. It consists of no step and exactly one transition; this path is called a sequence skip (right path Trans2).

a   ST: Step at the Top of the transition
      SB: Step Below the transition
b   SM: Step in the Middle

Table 4.14. (Continued)

We will see later that there is a textual representation only for the standard divergent sequence mode. It is not possible to write the user-defined priority and mutually exclusive modes in a textual notation (not portable).

IEC 61131-3 defines two types of errors to be avoided when programming networks in SFC. *Unsafe networks* allow an uncontrolled and uncoordinated activation of steps (outside of simultaneous sequences), see Example 4.49. *Unreachable networks* include some components that can never become active. Ideally, either the programming system should prevent the user from developing such networks or the run-time system should detect the existence of such networks and indicate an error.

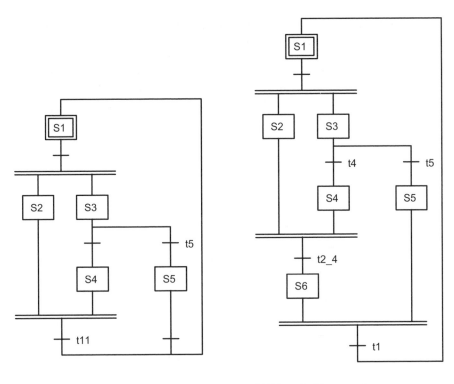

a) Unsafe network   b) Unreachable network

Example 4.49. Two networks, syntactically definable in SFC, which are, however, semantically erroneous and which must, therefore, be detected by the programming system and/or the runtime system of the PLC.

Assume that steps S2 and S3 of Example 4.49 a) have been activated. The transition t11 evaluates to FALSE (so S2 remains active), but t5 becomes TRUE; it is now possible by means of an S5 → S1 activation to restart S2 although S2 is already active. This is not allowed because it leads to an uncontrolled increase in the number of active tokens. No increase of active tokens outside of a simultaneous sequence is permitted. This network is **unsafe**.

In Example 4.49 b), by activating S2 and S3, the system may become deadlocked. The active token may move to S5 after an S3 → S5 transition. Now t1 is waiting for S6 to be activated. However, S6 will never be activated because t2_4 is waiting for S4 to be activated. This will never happen because t5 was activated instead of t4. Therefore, the terminating condition required for the active steps S5 **and** S6, in order to activate t1 is not **reachable**.

It would be easy to control the safety of networks if it were only necessary to ensure that no jumps were programmed out of a simultaneous sequence. However, it is also possible to develop safe or reachable networks with such critical jumps.

IEC 61131-3 describes a reduction algorithm to prevent these unwanted networks ([IEC 61131-8]). However, this algorithm is not foolproof; it is still possible to write safe and completely reachable networks which this algorithm forbids. (See Example 4.50).

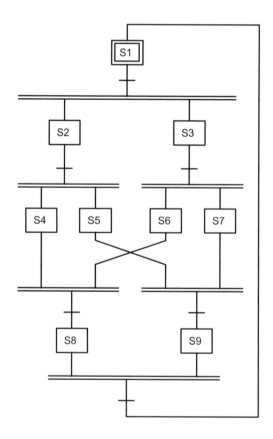

Example 4.50. Safe network, despite a jump out of a simultaneous sequence

4.6.3 Detailed description of steps and transitions

The two basic elements of an SFC network are:

1) Steps,
2) Transitions.

The *step* determines a set of instructions executed as long as the step is active.

The point of (de-)activation of steps is determined by the corresponding *transition* and defined by a *Boolean transition condition*. When a transition condition evaluates to TRUE, it stops the predecessor step that was previously active and activates the next successor step(s).

A transition condition may be programmed in IL, ST, LD or FBD. It must produce a Boolean value.

Step.
A step is represented graphically by a rectangular box.

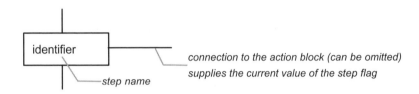

a) Graphical representation

STEP identifier:
        (* step body, for a description of actions see Example 4.55 *)
END_STEP

b) Textual representation

Figure 4.17. Step, represented by graphical (a) and textual notation (b)

The *step name* identifies the step. The *step flag* is an internally declared Boolean value returning the status of the corresponding step - active is TRUE and inactive is FALSE. This automatically declared variable, see Figure 4.18, may be used to control the instruction sequence of the action blocks or to observe the status from other steps.

A second implicitly declared variable shows the elapsed time since the activation of the step. If the step has never been active, its value is zero; if the step is no longer active, its value is the elapsed time of the last activation. This variable has the data type TIME.

step_name.X	step flag (a variable showing the status of the step)
step_name.T	elapsed time (a variable showing the activation time)

Figure 4.18. Information for a step: Activation status and time elapsed; the variables are declared implicitly and updated by the system.

To obtain the previous value of this time variable (for the last cycle) also in the event of a shutdown, the SFC structured POU must be an instance of type RETAIN.

The step name, step flag and the time variable are local to the POU. These variables are write-protected, i.e. their value cannot be changed directly by the user.

A step becomes active as soon as the condition of the predecessor transition connected above the step evaluates to TRUE. The step remains active until the condition of the successor transition connected at its bottom signals TRUE.

Be careful: Within an active step, it is possible to start an action with a time delay. Therefore, the situation could arise that the step is deactivated just as the delayed action starts. Now you have an inactive step with a running action!

Every SFC network has a special step, the *initial step*. Calling an SFC-structured POU for the first time means activating this initial step. If the POU consists of several SFC networks, every network starts via its own special initial step and – except for the user-programmed synchronisation points – runs afterwards independently of the other networks.

Any network step may be declared to be the initial step, as long as none of the rules in Section 4.6.2 concerning unsafe and unreachable networks are violated. If it has no predecessor in the network, the incoming line can be omitted.

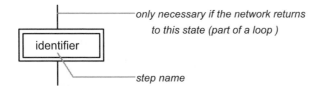

a) Graphical representation

INITIAL_STEP identifier:
      (* body of step; description of actions see Example 4.64 *)
END_STEP

b) Textual representation

Figure 4.19. Initial step, represented by graphical (a) and textual notation (b)

Transition.
In graphical representation, a *transition* consists of an incoming line (connection to predecessor steps), a transition condition and an outgoing line (connection to one or several successor steps). The transition condition is a Boolean expression.

A transition passes control to one or more successor steps if the following conditions apply:

1) All preceding steps that are directly connected to this transition are active.
2) The transition condition evaluates to TRUE.

Using the graphical representation, the transition condition can be specified by **three** methods. See also Figures 4.20, 4.21 and 4.22.

1)   The transition condition is written beside the transition (**immediate syntax**).
2)   **Connectors** connect the transition and transition condition.
3)   A **transition name** identifies the transition and transition condition.

Graphical representation:	Example:

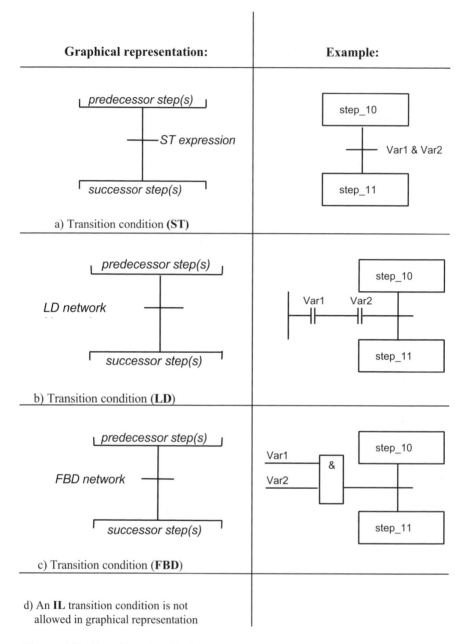

a) Transition condition (**ST**)

b) Transition condition (**LD**)

c) Transition condition (**FBD**)

d) An **IL** transition condition is not
allowed in graphical representation

Figure 4.20. Transition (graphical), syntax: ***immediate***. The transition condition is a
Boolean expression written in ST, LD or FBD (IL and SFC not allowed).

As shown in Figure Abb. 4.20, it is possible to write the transition condition immediately next to the transition. Another method is to use connectors. The Boolean value evaluated from a network is passed to the transition via a connector:

Graphical representation: | **Example:**

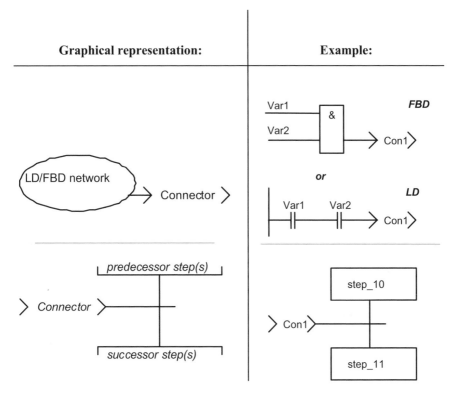

Figure 4.21. Transition (graphical), syntax: *connector*. The value of the network is passed to a transition (not allowed: IL and ST).

The third method uses the transition name to couple a transition condition, written independently of the SFC diagram, to the transition. Using this method, the transition condition can also be programmed in one of the textual languages. An additional advantage of this method is that one single transition condition can be specified for more than one transition if they all use the same evaluation expression; see Figure 4.22.

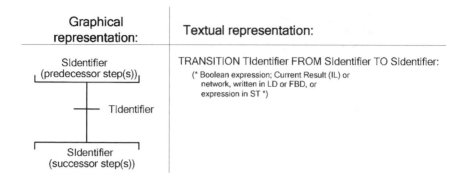

Figure 4.22. Transition (graphical/textual); representation with transition names; SIdentifier stands for step name, TIdentifier stands for transition name.

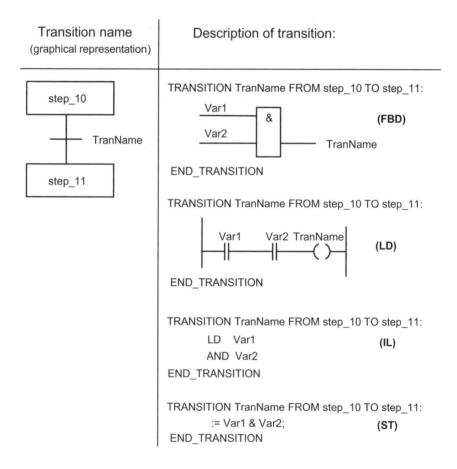

Example 4.51. Examples of transition conditions with transition names

How to use transition names is described in Example 4.51 for all four programming languages.

IEC 61131-3 defines another two methods for IL and ST where no transition names are used. It should be noted, however, that the definition of lines next to the graphical transition conditions is not allowed in IL. Therefore the programming system has to generate a transition name when the textual representation is recompiled to graphical representation.

```
TRANSITION FROM identifier TO identifier:
    (* IL instructions; last code line determines the Boolean condition: *)
    (* Current Result *)
END_TRANSITION
```

Example:
```
TRANSITION FROM step_10 TO step_11 :
    LD      Var1
    AND     Var2       (* CR with a Boolean value *)
END_TRANSITION
```

Figure 4.23. Textual description of a transition condition written in IL. It is possible to define more than one step identifier after the "FROM" or "TO" parts (written in parentheses, separated by commas).

```
TRANSITION FROM identifier TO identifier :=
        (* ST expression *)
END_TRANSITION
```

Example 1:
```
TRANSITION FROM step_10 TO step_11
    := Var1 & Var2;
END_TRANSITION
```

Example 2 defining more than one predecessor and successor steps:
```
TRANSITION FROM (S_1, S_2, S_3) TO (S_4, S_5)
    := Var1 & Var2;
END_TRANSITION
```

Figure 4.24. Textual transition condition in ST. It is possible to define more than one step identifier (written in parentheses, separated by commas).

Within a block for determining the transition condition, only Boolean expressions may be used and no values may be assigned to variables other than the transition name. This means also that calls of FB instances are not allowed because they could contain value assignments. Calling a function is allowed.

It can be concluded from the above that transitions are not suitable for calculating and storing values and for calling FB instances. The actions associated with each step are intended for use with these operations.

4.6.4 Step execution using action blocks and actions

When a step is active, the instructions associated with the step have to be executed.

These instructions are written in so-called action blocks. An *action block* describes a set of *actions* which have to be executed when the step becomes active. Each action can be associated with a number of special conditions called *qualifiers* that indicate how and when the execution of the actions is to be started, i.e. immediate start, with delay, on a rising edge etc.

Simple action blocks can be used which only set or reset Boolean variables *(Boolean actions)* for process synchronisation; see Example 4.52. Complex action blocks contain or refer to instruction sequences or networks, programmed in an IEC 61131-3 language; see Example 4.53.

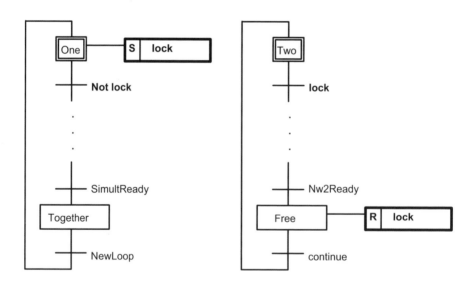

Example 4.52. Two SFC networks synchronised by the Boolean action lock using the qualifier **S**(et)/**R**(eset) (Set sets lock to TRUE, Reset sets lock to FALSE).

When this POU is called, the network on the left-hand side sets the Boolean variable lock to TRUE. Therefore, the following transition does not "fire" and step One remains active. The right network is started when lock becomes TRUE. When the step Free becomes active, it resets the variable lock to FALSE and the left network can then continue. The Boolean "lock" is used to synchronise the two networks.

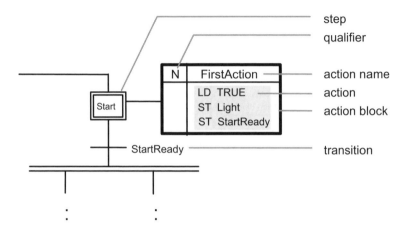

Example 4.53. Extract from the SFC network of Example 4.48, extended with a (**non-**Boolean) action block for step Start.

The action block of Example 4.53 defines the activities occurring after the activation of step Start. TRUE is written to the variables Light and StartReady. In this example, StartReady defines the transition condition of this step. This means that the step instructions are not executed cyclically, but the step is immediately deactivated and the successor steps are activated.

An *SFC structured network* reflects the following hierarchy:

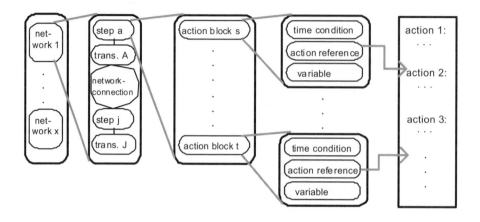

Figure 4.25. Hierarchy of an SFC structured POU

4.6.5 Detailed description of actions and action blocks

Actions.
Actions can be used to:

1) Define the instructions of a step of an SFC network **or**
2) Define a sequence of instructions in an LD or FBD network that are executed under certain conditions.

Most programming systems implement only the first variant.

An action consists of:

1) An assignment to a Boolean variable (*Boolean action*) **or**
2) An instruction sequence (*action or instruction action*), programmed in IL, ST, LD, FBD or as an SFC structured network.

A Boolean action is the assignment of a value to a variable (declared in a VAR or VAR_OUTPUT declaration block).

The statements of an instruction action are executed repeatedly as long as the associated step is active.

Action block.

An *action block* consists of an action together with the execution condition (called the action qualifier) and a Boolean *"indicator" variable*. Action blocks are defined in both graphical and textual notation; see Figure 4.26.

a) Simple graphical action block

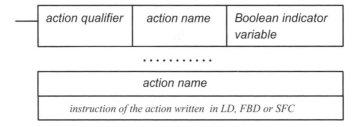

b) Graphical action block and attached instruction block

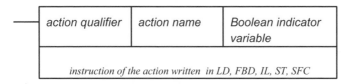

c) Graphical action block and separate instruction block

action name (action qualifier possibly with parameters, indicator var.)
...

ACTION *action name*:

(* *instruction of the action written in IL or ST* *)

END_ACTION

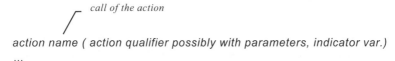

d) Textual action block and separate instruction block

Figure 4.26. Four methods of defining an action block / actions

There are four notations to describe and call an action block or actions as described in Figure 4.26:

a) The execution condition (action qualifier) and action name together make up a simple action block.
b) Similar to a), extended with an optional indicator variable and an instruction block programmed in LD, FBD, IL, ST or SFC.
c) Equivalent representation to b). However, the action is not described within the block but in another part of the program (its representation is in LD, FBD or SFC). These two information blocks are linked via the action name.
d) Textual representation of c). Action name and qualifier with optional parameters are written within an (INITIAL) STEP...END_STEP block. It is not required that the ACTION...END_ACTION textual block appears directly after the action call

It is possible to program a step with no action block (do nothing) or with one or more action blocks, see Examples 4.54 and 4.55. However, the entire block of action blocks has only one common (activity) connection to the step (*step flag*) (.

The system monitors the execution conditions of all step action blocks between activation and de-activation of the step. This execution condition, called an *action qualifier* controls

-    The execution time of the associated instructions **or**
-    The value of the specified variable (Boolean action).

Action qualifiers are explained in Section 4.6.7.

The *action name* defines an action. The scope of this action name is the POU. An action name either specifies a Boolean assignment or references an action with instructions. The execution time of an action is defined by one or more action blocks (of different steps) all of which reference the same common action.

The *indicator variable* is a special variable written to within an action to reflect the status of the action (for example the value TRUE might indicate "action is active"). This variable must be declared just like any other variable. The idea of the indicator variable is to show the user the status of a step outside of its associated action description. Mostly this syntax is used either when the action is programmed textually and not written directly after the call, or the action is an SFC structure. The assignment to this variable must be programmed manually; there is no automatic system administration of the indicator variable (most systems have not implemented the use of the indicator variable).

The *instruction part* of an action block defines a sequence of instructions.

The indicator variable and the instruction part are optional. Without an indicator variable, the action name can be used to obtain information about the state of the step. It is mandatory to declare an action name as a Boolean variable if this name has no defined action block of the same name and therefore must be interpreted as a Boolean variable.

The following general rules apply:

1) Every step (including associated actions) is executed at least once after it has been activated.
2) After deactivation, the step and all actions are called one last time to ensure well-defined step termination, i.e. to stop timers, reset variables etc. The step flag is TRUE for each execution of the step except this last execution.

Some programming systems display networks consisting only of steps and transitions. Information on the assigned action blocks can be displayed, if required (e.g. by double-clicking in a separate window). This is also consistent with the Standard.

4.6.6 Relationship between step, transition, action and action block

The following examples summarise the terms and illustrate the use of SFC definitions. They are followed by a description of the action qualifier.

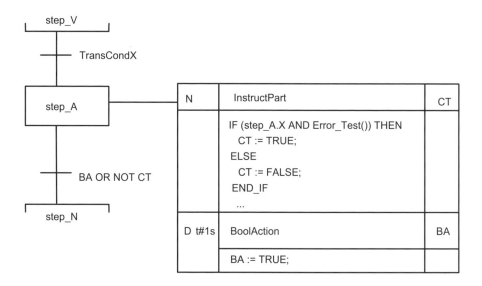

Example 4.54. Two actions: BoolAction is an action that sets variable BA after a time delay. InstrucPart is an instruction action which contains some instructions.

```
TRANSITION FROM step_V TO step_A := TransCondX;
END_TRANSITION
TRANSITION FROM step_A TO step_N := BA OR NOT CT;
END_TRANSITION
STEP step_A:
        InstrucPart(N, CT);
        BoolAction(D, t#1s, BA);
END_STEP
ACTION InstrucPart :
        IF (step_A.X AND Error_Test()) = TRUE THEN
                CT := TRUE;
        ELSE
                CT := FALSE;
        ...
END_ACTION
ACTION BoolAction :
        BA := TRUE;
END_ACTION
```

Example 4.55. Example 4.54 in textual notation

As soon as the transition condition TransCondX in Example 4.54 becomes TRUE ("fires"), step_A becomes active. Now, the action qualifiers of this step are evaluated.

The qualifier "N" of the first action block, InstrucPart, indicates that the step flag is used to control the execution of the action. The associated instruction part will be executed as long as the step is active.

step_A is active as long as no error occurs (ErrorTest returns TRUE) and BA is FALSE. If one of the two conditions changes status, the transition becomes TRUE and step_A receives the status "inactive". InstruPart is called one last time and CT is set to FALSE (the AND expression evaluates to FALSE because step_A.X is now FALSE).

D means "Delay"; after a delay of the specified time (1 second in the example), the variable BA is set to TRUE. For a Boolean action, it is possible to omit the variable name. In that case, a variable with the same name as the action name takes over the job of the indicator variable. In our example, it would be possible to use a Boolean variable, BoolAction, instead of BA, to program the transition condition.

The instruction part of Example 4.55 is written in ST. An equivalent language construct could also be written in IL or an equivalent network drawn in LD or FBD. Actions can also be programmed as an SFC structure.

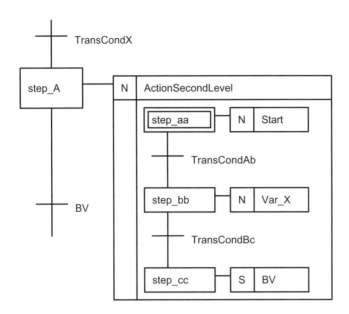

Example 4.56. Substructure of an action written in SFC (nested SFC structure). The indicator variable BV in step_cc is set in the second level network and influences the deactivation of step step_A.

Example 4.56 describes an action block with a sublevel SFC structure.

SFC networks of a lower level, like the network ActionSecondLevel in Example 4.56, are executed as long as the calling action / step (step_A) is active. Deactivation of the calling step stops the scheduling of the sub-networks. In our example, the variable BV ensures that the whole sub-network ends before the step_cc is deactivated. BV is part of the transition condition of step_A.

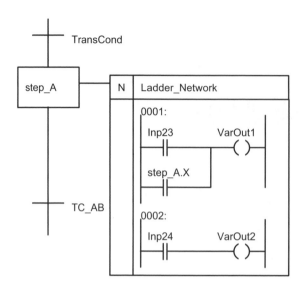

Example 4.57. Instruction action graphically programmed. The step flag of step_A is used to allow separate reaction during the last execution after deactivation.

As soon as step_A in Example 4.57 is active, VarOut1 is TRUE because the step flag evaluates to TRUE. It is possible to program more than one network in an instruction action. As soon as TC_AB deactivates step_A, the action is called for the last time. Now Inp23 is assigned to VarOut1 because step_A.X evaluates to FALSE.

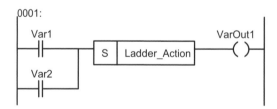

Example 4.58. Programming an action block **within** an LD network

Unlike Example 4.57 where the action block is assigned to a step and programmed with LD, the *action block* of Example 4.58 is programmed directly within an LD network. In this case, the Boolean variable, Ladder_Action, is set to TRUE when either Var1 or Var2 evaluate to TRUE.

An action block like Example 4.58 can also be programmed in an FBD network. The input of an action block is connected to an FBD output.

Example 4.59. Programming an action block **within** an FBD network

The left connection of an action block triggers the action just like the step flag in an SFC. The right-hand line is assigned the value of the left input line. It is not possible to convert this notation to the textual languages, IL and ST.

4.6.7 Action qualifiers and execution control

The following section discusses the control flow within an SFC structured POU.

Qualifier.
The execution time of an action depends on the step flag and the *action qualifier*. The following letters are valid for qualifiers:

Qual	Short term	Explanation
(empty)		The same as qualifier N.
N	Non-stored	Current value of the step flag
R	Overriding Reset	Reset to FALSE, stored
S	Set	Set to TRUE, stored
L	Time Limited	Evaluate to TRUE until the time expires or the step is deactivated
D	Time Delayed	Evaluate to TRUE after the specified time duration until the deactivation of the step
P	Pulse	Evaluate to TRUE on a FALSE → TRUE transition (edge detection).
SD	(unconditional) Stored and Time Delayed	Evaluate to TRUE after the specified time duration independently of the step deactivation
DS	Time Delayed, Stored	Evaluate to TRUE after the specified time duration and the step is still active
SL	Stored and Time Limited	Evaluate to TRUE until the end of the specified time duration
P1	Pulse, rising edge	The action block is executed once after step activation (see also P0: step deactivation)
P0	Pulse, falling edge	The action block is executed once after step deactivation (see also P1: step activation)

Table 4.15. Qualifiers (Qual) in SFC

For a detailed graphical description see Example 4.62.

Every action is controlled by an action-specific scheduler called *action control*. All action blocks calling the action influence its execution time. The action control and its Q output are invisible to the user in most programming systems.

The action control of every action, mostly implemented as a system function block, is part of the (SFC specific) operating system and determines the start and stop condition of an action. The inputs of a special action control are connected (invisibly to the programmer) to the action blocks (step flags, qualifiers) which use this action. The inputs are evaluated consistently according to the rules in Figure 4.28. Output Q is a Boolean value and shows the start / stop condition of the action. The possibility of programming this variable explicitly is implementation-dependent.

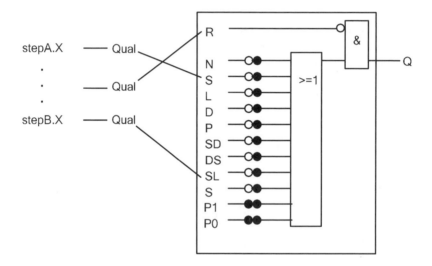

Figure 4.27. Action control of an action. All action blocks calling this action influence the inputs by their step flags and qualifiers (Qual). The inputs are computed in accordance with the rules below (represented schematically by "●●"). The result (Q output) called *action flag* decides whether the action is executed (TRUE) or not (FALSE)

Every action block has an invisible connection to the input parameter identified by the qualifier of the action control for this action. So the relevant Qual input receives the value of the step flag of all action blocks using this action which have specified the use of that qualifier (see Example 4.60). The outputs of all qualifier expressions are wired by OR with one exception: RESET, with the highest priority, is wired by AND with the other results.

The result of the OR and AND operation is written to a variable called *Q output*.

The value of the Q output

- Is assigned to the action variable (Boolean action),
- Controls the execution of instruction actions. As long as the Q output evaluates to TRUE, the instructions (network) are executed cyclically; otherwise, they are not executed.

The functions of the various qualifiers are illustrated in Figure 4.28.

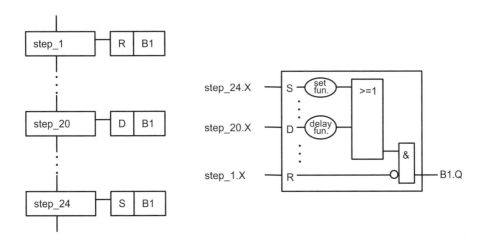

Example 4.60. Action control of action B1 called by three action blocks. The qualifiers of these action blocks and their step flags control the value of B1.Q.

To understand the functionality of the action control in Figure 4.28, imagine an action X used by only one action block with qualifier "N". When the associated step is activated, the step flag evaluates to TRUE, which means that the "N" input of the action control is also TRUE. The OR function of the action control generates TRUE. Because no RESET is used for this action, the negated RESET input is also TRUE. The result of the AND operation becomes TRUE and so the Q output evaluates to TRUE. The action is executed until the step flag evaluates to FALSE.

It is forbidden to control an action by two qualifiers with time parameters (L, D, SD, DS or SL) because SFC does not define any precedence rule for these.

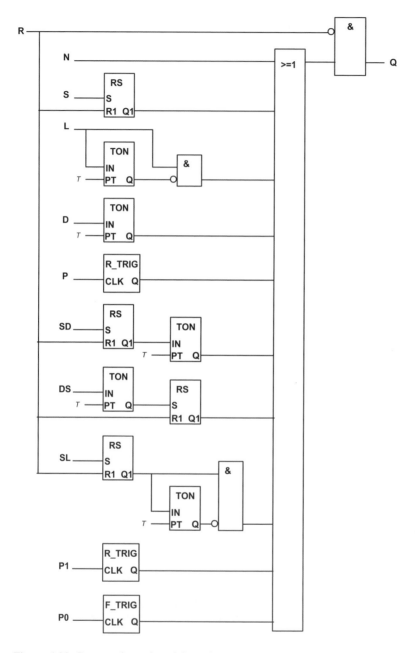

Figure 4.28. Computation rules of the action control of *one* action. All active steps calling this action send TRUE to the equivalent input (N, R, S, L, D, P, SD, DS, SL, P1, P0). T is a parameter of type TIME specified in the action block. The 12 FB instances correspond to the IEC 61131-3 standard FBs (without FB instance name). This control algorithm is part of the SFC operating system.

The system provides the action flag (Q output) in Example 4.60 as Boolean variable. Edition 2 of the Standard defines another Boolean variable: the *activation flag*. It remains active as long as the action is executed, i.e. by one execution loop longer than the action flag that evaluates to FALSE at the last execution cycle (after the associated step has been deactivated).

If the activation flag is used in a system, action control in Figure 4.28 has to be extended as shown in the following illustration.

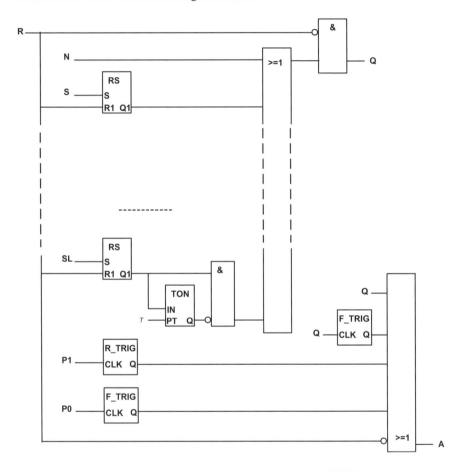

Figure 4.29. Extension of Figure 4.28 by the activation flag; TRUE indicates that the corresponding action is currently active (possibly including an already inactive step).

The following examples demonstrate how the value of the Q output is dependent on the qualifier.

The Boolean variable A1 in the examples is influenced by the qualifier value and the output Q of the action control. The examples demonstrate the time behaviour between these variables.

In the case of an instruction action, the corresponding instruction part is executed cyclically as long as the operating system computes the Q output of A1 to be TRUE.

To demonstrate the mutual influence of the qualifiers of several action blocks, a second step SX, which also controls A1, is introduced.

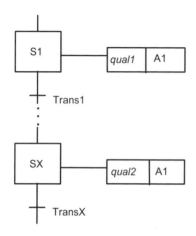

Example 4.61. Step and action block description for Example 4.62.

qual1	qual2	Example
N		Direct copy of the step flag: S1 ⎍⎍ A1 ⎍⎍
S	R	Set and Reset of an action: S1 ⎍⎍ SX ⎍⎍ A1 ⎍⎍

Example 4.62. Value of the Q output of A1 (action is executed, or assignment to a Boolean variable), influenced by a qualifier of one action block (qual1) calling this action, or of two action blocks qual1 and qual2. (Continued on next page)

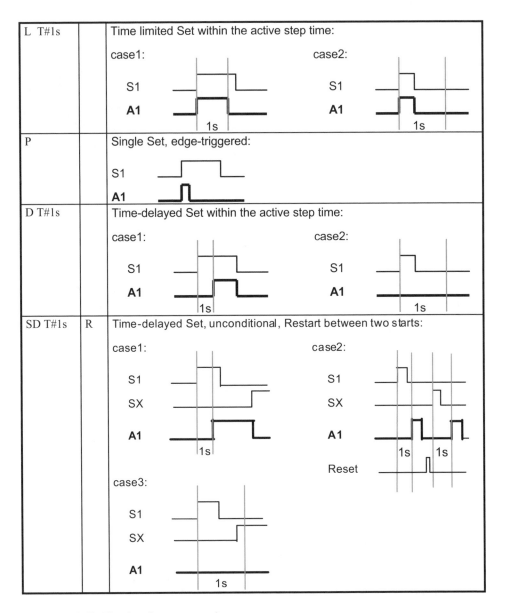

Example 4.62. (Continued on next page)

The difference between SD and DS is the influence of the step flag: DS starts the action after the delay time only when the step flag is still TRUE. With the qualifier SD the action is executed unconditionally (until a RESET is scheduled for this action).

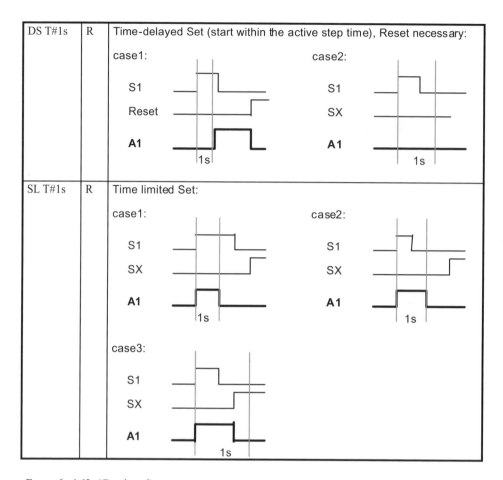

Example 4.62. (Continued)

Sequential control.
In the programming languages IL or ST, a sequence of instructions is executed one instruction after another. The sequence can be interrupted using instructions like CAL or JMP. The method of controlling the execution sequence is different in SFC:

An SFC structured network is divided into steps and transitions which are evaluated and executed cyclically. A step always has either active or inactive status and all transitions are evaluated every cycle to either TRUE or FALSE. The active step list for the pending cycle is dependent on the TRUE / FALSE evaluation of the transitions on which the steps depend.

The execution of all instructions in an SFC network is scheduled by the PLC / SFC operating system using the following algorithm:
Note: Activate a step means: execute all instructions of this step.

1) Activate initial step (only at the first call). Otherwise:
   Deactivate all active steps whose successor transitions are TRUE and activate all steps following these transitions.
2) Check the Q output of all action controls. If a TRUE ? FALSE edge has just been detected, execute the associated action one last time (this is the last activation).
3) Execute all actions whose action control evaluates to TRUE (Q output).
4) As soon as the instructions and / or the Boolean actions are finished, process all inputs and outputs by reading and writing the values of the variables to/from the physical I/Os (directly declared variables; see Section 3.5.1).
5) Evaluate the transitions and continue with step 1).

There is no sequential processing of instructions in the networks, but a continuous evaluation of the status of steps and transitions.

Consequently, an SFC network or POU has no explicit end, or RETURN. If there is no subsequent transition, the program does not jump back automatically to the *initial step* (not even in case of a substructure in SFC as shown in 4.56), the POU remains in the last active step.

The difficulty when calling an SFC-structured POU by a non-SFC-structured POU is to define the point of time when the output parameters become active because the SFC structure is always active (no RETURN). A solution could be, for example, to include a "release operand" for the called POU. Moreover, the system reaction must be defined because calling an SFC-structured POU a second time would otherwise result in an unsafe network or further instantiation.

The behaviour has not been standardised to date and is solved in different ways by the different programming systems. For example, if SFC structuring is blocked for function blocks (i.e. allowed only for PROGRAM), the inconsistency addressed above does not apply.

As a rule, the algorithm for sequence control implementation uses a (global) data structure which includes the step states and transitions of all POUs ever called. Thus, all activated SFC networks are evaluated simultaneously.

Another method is the (local) status and transition monitoring for each POU: if an FB is called, all currently active transitions and states of the FB are evaluated and executed. Control then returns to the calling program (PROGRAM). The program ends with the exchange of variables (*End of cycle*). This ensures that in each cycle only those transitions and states are monitored and executed which belong to the POUs called in the relevant cycle.

4.6.8 Example: "Dino Park"

All theme parks offer a variety of buildings and animal enclosures. Since the film Jurassic Park, it is a well-known fact that people no longer stroll around on foot in this high-tech world. The visitors are driven to the various attractions in comfortable vehicles that are linked together. Because the visitors in our theme park "*Dino Park*"pay well for the privilege of entering the park, only one group of people is in the park at any one time. The members of the group enter the ticket office together and climb into the cars. The adventure starts and the visitors are driven automatically to the first dinosaur enclosure.

The park consists of 6 areas / attractions (Att). The electronics have to control each of these attractions with the following default specification:

Att1: Ticket office. Turn on the main power via the Boolean actuator ElectOn.
An elevator is moved down by the variable ElevDown to a loading point (for food in Att5). The stopping of the elevator is controlled by a hardware sensor positioned to stop the elevator at the right place.

Att2: Turn on the waterfall as long as visitors remain in the attraction's building.
An action with instructions controls the mechanism of dinosaur puppets.
In order to have sufficient compressed air for subsequent attractions in other buildings, another action starts up the compressed air generator.
After 10 minutes, the sound machine is started, which generates dinosaur screams, which are the signal for departure from this building.

The group can now separate into two halves: Intrepid visitors are in for a watery surprise in Att4, while the others are treated to a more harmless version in Att3.

Att3: Turn on lights while the visitors are in the building.
Using compressed air, confetti is blown into the room. As soon as a visitor finds and activates the switch ActionStop, the action in Att3 and Att4 stops and the ride continues.

Att4: This is like Att3 with the exception that the confetti is replaced by water!

Att5: Outdoor enclosure.
Feeding time for the dinosaurs: The elevator with food is started. It takes 15 seconds until it reaches its top position.
The compressed air generator is terminated.
After 20 minutes, a money sensor is activated. If sufficient money is inserted by the visitors by the end of this time, the ride continues from the beginning (back to Att1).

Att6: End of attraction. Turn out the light.

```
PROGRAM Dinopark
VAR
        ElectOn             AT %QB0.0 : BOOL;
        ElevDown            AT %QB1.0 : BOOL;
        ElevUp              AT %QB1.1 : BOOL;
        Waterfall           AT %QB2.0 : BOOL;
        Scream              AT %QB3.0 : BOOL;
        Pressure            AT %QB4.0 : BOOL;
        Light3On            AT %QB5.3 : BOOL;
        Light4On            AT %QB5.4 : BOOL;
        WaterGo             AT %QB6.0 : BOOL;
        ConfettiGo          AT %QB6.1 : BOOL;
        ActionStop          AT %IB1.0  : BOOL;
        Money               AT %IW2    : INT;
        MoneyStatus                    : BOOL;
END_VAR
```

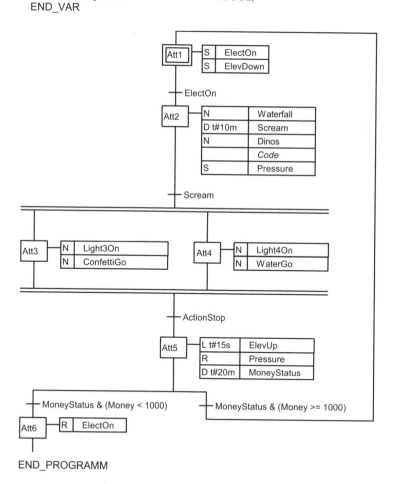

END_PROGRAMM

Example 4.63. SFC structured POU: Dino Park

```
VAR ...
END_VAR

INITIAL_STEP Att1:
      ElectOn(S);
      ElevDown(S);
END_STEP

STEP Att2:
      Waterfall(N);
      Scream(D, t#10m);
      Dinos(N);
      Pressure(S);
END_STEP
(* additional step declarations *)
...
TRANSITION FROM Att1 TO Att2 := ElectOn; END_TRANSITION
TRANSITION FROM Att2 TO (Att3, Att4) := Scream; END_TRANSITION
TRANSITION FROM (Att3, Att4) TO Att5 := ActionStop; END_TRANSITION
TRANSITION FROM Att5 TO Att1 := MoneyStatus & (Money >= 1000);
END_TRANSITION
TRANSITION FROM Att5 TO Att6 := MoneyStatus & (Money < 1000); END_TRANSITION

ACTION Dinos :
      (* instructions *);
END_ACTION
```

Example 4.64. Example 4.63 in textual notation (extract)

Comments on the network for the dinosaur park

Step	Comment
Att1	Initial step; this step is activated when the program is called.

Att1: Set the Boolean variables ElectOn and ElevDown to TRUE, as soon as the step flag of Att1 evaluates to TRUE. Additionally ElectOn is used to trigger the successor transition. Att6 will reset this Boolean action.

Att2: The Boolean variable Waterfall is set to TRUE (to keep running the waterfall) as long as the step flag of Att2 is active.

The variable Scream is set to TRUE after a delay time of 10 minutes.

While Att2 is active, the instructions of action Dinos are executed cyclically.

The variable Pressure is set to TRUE.

Att3 / Att4: After the trigger of Scream (FALSE -> TRUE) both steps are active simultaneously. All Boolean variables of the action blocks become TRUE.

As soon as the sensor ActionStop reports TRUE, both actions are deactivated and Att5 becomes active.

Att5: Now ElevUp is assigned TRUE for 15 seconds and then FALSE.

The variable MoneyStatus is set to TRUE after a delay time of 20 minutes.

The two subsequent transitions are mutually exclusive by the test using variable MoneyStatus. Only one transition can become true at any given time.

Att6: The program proceeds on the assumption that the POU is called only once. Consequently, this POU remains in the step Att6 (if no more money is put in).

The Boolean variable ElectOn started in the initial step is reset.

5 Standardised PLC Functionality

The IEC not only standardises the syntax of programming languages, but even goes a step further to unify the implementation of typical PLC functions, such as timers, counters or special arithmetic operations.

The standard does this by defining typical PLC functions and function blocks and describing their behaviour exactly. These elements are known as *standard functions* and *standard function blocks* respectively. Their names are reserved keywords.

If the functions and function blocks in the programming systems and block libraries of different manufacturers are given the names specified in the standard, they must comply with the rules set out in the standard. Manufacturers can also offer additional PLC functions which, for example, support particular hardware properties or other characteristics of a PLC system.

The definition of an unambiguous standard for PLC functions is an essential requirement for uniform, manufacturer- and project-independent training, programming and documentation.

This chapter gives an overview of the most important standard functions and function blocks as well as the concepts used within them:

1) **Standard functions (std. FUN)**
   - Calling interface
   - Extensibility
   - Overloading
   - Examples
2) **Standard function blocks (std. FB)**
   - Calling interface
   - Examples

The standard functions correspond to the basic logical operators used in conventional PLC systems (addition, bit-shifting, comparison etc.), whereas the standard function blocks are responsible for PLC functions with status information, such as timers, counters, R/S flip-flops and edge detectors.

K.-H. John, M. Tiegelkamp, *IEC 61131-3: Programming Industrial Automation Systems*, 2nd ed., DOI 10.1007/978-3-642-12015-2_5,
© Springer-Verlag Berlin Heidelberg 2010

In the following sections, the calling interfaces (input and output variables and the function [return] value) for standard functions and function blocks are described in detail. Practical examples accompany the various descriptions.

The graphical declarations of all standard functions and function blocks, together with a short functional description, are given in Appendices A and B.

The general usage of functions and function blocks has already been discussed in Chapter 2 and the properties of their formal parameters in Chapter 3.

5.1 Standard Functions

IEC 61131-3 defines the following eight groups of standard functions:

1) Data type conversion functions,
2) Numerical functions,
3) Arithmetic functions,
4) Bit-string functions (bit-shift and bitwise Boolean functions),
5) Selection and comparison functions,
6) Character string functions,
7) Functions for time data types,
8) Functions for enumerated data types.

Table 5.1 summarises all standard functions of the standard. The special functions for time data types (ADD, SUB, MUL, DIV, CONCAT) and enumerated data types (SEL, MUX, EQ, NE) are grouped together with the other functions in the categories Arithmetic, Comparison, Selection and Character string.

The table gives the function name, the data type of the function values and a short description of the function. Together with Table 5.2, it also gives the names and data types of the input variables.

With the exception of the generic data types, the abbreviations for the data types of the input variables and function values in Table 5.1 are listed in Table 5.2. These abbreviations correspond to the names of the input variables used by IEC 61131-3 for the respective standard functions. ENUM is an additional abbreviation used to make Table 5.1 clearer.

Standard functions (with data types of input variables)		Data type of function value	Short description	over-loa-ded	exten-sible
Type conversion					
* _TO_**	(ANY)	ANY	Data type conversion	yes	no
TRUNC	(ANY_REAL)	ANY_INT	Rounding up/down	yes	no
BCD_TO_**	(ANY_BIT)	ANY	Conversion from BCD	yes	no
* _TO_BCD	(ANY_INT)	ANY_BIT	Conversion to BCD	yes	no
DATE_AND_TIME_TO_- TIME_OF_DAY	(DT)	TOD	Conversion to time-of-day	no	no
DATE_AND_TIME_TO_- DATE	(DT)	DATE	Conversion to date	no	no
Numerical					
ABS	(ANY_NUM)	ANY_NUM	Absolute number	yes	no
SQRT	(ANY_REAL)	ANY_REAL	Square root (base 2)	yes	no
LN	(ANY_REAL)	ANY_REAL	Natural logarithm	yes	no
LOG	(ANY_REAL)	ANY_REAL	Logarithm to base 10	yes	no
EXP	(ANY_REAL)	ANY_REAL	Exponentiation	yes	no
SIN	(ANY_REAL)	ANY_REAL	Sine	yes	no
COS	(ANY_REAL)	ANY_REAL	Cosine	yes	no
TAN	(ANY_REAL)	ANY_REAL	Tangent	yes	no
ASIN	(ANY_REAL)	ANY_REAL	Arc sine	yes	no
ACOS	(ANY_REAL)	ANY_REAL	Arc cosine	yes	no
ATAN	(ANY_REAL)	ANY_REAL	Arc tangent	yes	no
Arithmetic *(IN1, IN2)*					
ADD {+}	(ANY_NUM, ANY_NUM)	ANY_NUM	Addition	yes	yes
ADD {+} [a]	(TIME, TIME)	TIME	Time addition	yes	no
ADD {+} [a]	(TOD, TIME)	TOD	Time-of-day addition	yes	no
ADD {+} [a]	(DT, TIME)	DT	Date addition	yes	no
MUL {*}	(ANY_NUM, ANY_NUM)	ANY_NUM	Multiplication	yes	yes
MUL {*} [a]	(TIME, ANY_NUM)	TIME	Time multiplication	yes	no
SUB {-}	(ANY_NUM, ANY_NUM)	ANY_NUM	Subtraction	yes	no
SUB {-} [a]	(TIME, TIME)	TIME	Time subtraction	yes	no
SUB {-} [a]	(DATE, DATE)	TIME	Date subtraction	yes	no
SUB {-} [a]	(TOD, TIME)	TOD	Time-of-day subtraction	yes	no
SUB {-} [a]	(TOD, TOD)	TIME	Time-of-day subtraction	yes	no
SUB {-} [a]	(DT, TIME)	DT	Date and time subtraction	yes	no
SUB {-} [a]	(DT, DT)	TIME	Date and time subtraction	yes	no
DIV {/}	(ANY_NUM, ANY_NUM)	ANY_NUM	Division	yes	no
DIV {/} [a]	(TIME, ANY_NUM)	TIME	Time division	yes	no
MOD	(ANY_NUM, ANY_NUM)	ANY_NUM	Remainder (modulo)	yes	no
EXPT {**}	(ANY_NUM, ANY_NUM)	ANY_NUM	Exponent	yes	no
MOVE {:=}	(ANY_NUM, ANY_NUM)	ANY_NUM	Assignment	yes	no

a   Special function for time data type. The generic input and output data type for ADD and SUB is therefore ANY_MAGNITUDE, see also Section 3.4.3

Table 5.1. Overview of the standard functions (continued on next page)

Standard functions (with data types of input variables)		Data type of function value	Short description	over-loa-ded	exten-sible
Bit-shift	**_(IN1, N)_**				
SHL	(ANY_BIT, ANY_INT)	ANY_BIT	Shift left	yes	no
SHR	(ANY_BIT, ANY_INT)	ANY_BIT	Shift right	yes	no
ROR	(ANY_BIT, ANY_INT)	ANY_BIT	Rotate right	yes	no
ROL	(ANY_BIT, ANY_INT)	ANY_BIT	Rotate left	yes	no
Bitwise	**_(IN1, IN2)_**				
AND {&}	(ANY_BIT, ANY_BIT)	ANY_BIT	Bitwise AND	yes	yes
OR {>=1}	(ANY_BIT, ANY_BIT)	ANY_BIT	Bitwise OR	yes	yes
XOR {=2k+1}	(ANY_BIT, ANY_BIT)	ANY_BIT	Bitwise EXOR	yes	yes
NOT	(ANY_BIT, ANY_BIT)	ANY_BIT	Bitwise inverting	yes	no
Selection	**_(IN1, IN2)_**				
SEL	(G, ANY, ANY)	ANY	Binary selection (1 of 2)	yes	no
SEL [b]	(G, ENUM, ENUM)	ENUM	Binary selection (1 of 2)	no	no
MAX [c]	(ANY_E, ANY_E)	ANY_E	Maximum	yes	yes
MIN [c]	(ANY_E, ANY_E)	ANY_E	Minimum	yes	yes
LIMIT [c]	(MN, ANY_E, MX)	ANY_E	Limitation	yes	no
MUX [c]	(K, ANY, ..., ANY)	ANY	Multiplexer (select 1 of N)	yes	yes
MUX [b]	(K, ENUM, ..., ENUM)	ENUM	Multiplexer (select 1 of N)	no	no
Comparison	**_(IN1, IN2)_**				
GT {>}	(ANY, ANY)	BOOL	Greater than	yes	yes
GE {>=}	(ANY, ANY)	BOOL	Greater than or equal to	yes	yes
EQ {=}	(ANY, ANY)	BOOL	Equal to	yes	yes
EQ {=} [b]	(ENUM, ENUM)	BOOL	Equal to	no	no
LT {<}	(ANY, ANY)	BOOL	Less than	yes	yes
LE {<=}	(ANY, ANY)	BOOL	Less than or equal to	yes	yes
NE {<>}	(ANY, ANY)	BOOL	Not equal to	yes	no
NE {<>} [b]	(ENUM, ENUM)	BOOL	Not equal to	no	no
Character string	**_(IN1, IN2)_**				
LEN	(STRING)	INT	Length of string	no	no
LEFT	(STRING, L)	STRING	string "left of"	yes	no
RIGHT	(STRING, L)	STRING	string "right of"	yes	no
MID	(STRING, L, P)	STRING	string "from the middle"	yes	no
CONCAT	(STRING, STRING)	STRING	Concatenation	no	yes
CONCAT [a]	(DATE, TOD)	DT	Time concatenation	no	no
INSERT	(STRING, STRING, P)	STRING	Insertion (into)	yes	yes
DELETE	(STRING, L, P)	STRING	Deletion (within)	yes	yes
REPLACE	(STRING, STRING, L, P)	STRING	Replacement (within)	yes	yes
FIND	(STRING, STRING)	INT	Find position	yes	yes

a   Special function for time data type
b   Special function for enumeration data type
c   ANY_E is the abbreviation of ANY_ELEMENTARY

Table 5.1. (Continued)

Input	Meaning	Data type
N	Number of bits to be shifted	ANY_INT
L	Left position within character string	ANY_INT
P	Position within character string	ANY_INT
G	Selection out of 2 inputs (gate)	BOOL
K	Selection out of n inputs	ANY_INT
MN	Minimum value for limitation	ANY_ELEMENTARY
MX	Maximum value for limitation	ANY_ELEMENTARY
ENUM	Data type of enumeration	

Table 5.2. Abbreviations and meanings of the input variables in Table 5.1

The function names are listed on the left-hand side of column 1 in Table 5.1. The meanings of the asterisks (*) in the function names of the "Type conversion" group are as follows (see also Appendix A):

* Data type of the input variable (right-hand side of column 1)
** Data type of the function value (column 2)

The names of the input variables of a function, if any, are given in the italic heading of the group, if they apply to the group as a whole. For example, the input variables for arithmetic functions are named IN1, IN2 and, when extended, IN3, IN4, ... If a function only has a single input variable, this does not have a name.

If a function with several inputs has only one input of same data type (overloaded), its variable name is "IN" (without number). This applies to LIMIT, LEFT, RIGHT, MID and DELETE.

Within IEC 61131-3, SEL is an exception to this uniformity. The inputs of this function are called G, IN0 and IN1 (instead of IN1, IN2).

Some standard functions have an alternative function name consisting of symbols in their graphical representation, as shown in curved brackets directly after the function name in Table 5.1. For example, the addition function can be called as ADD (operator in IL) or as "+" (graphical symbol in LD/FBD or within ST expressions).

5.1.1 Overloaded and extensible functions

The data types of the input variables are given in round brackets next to the function names in Table 5.1. Here generic data types, already introduced in Table 3.9, are also given for reasons of clarity. Each function whose input variable is described using a generic data type is called *overloaded* and has a "yes" in the corresponding column in Table 5.1. This simply means that the function is not restricted to a single data type for its input variables, but can be applied to different data types.

The data type of the function value (2nd column) is normally the same as the data type of its inputs. Exceptions are functions such as LEN, which expects a character string as its input but returns INT as its function value.

If a standard function can have a variable number of inputs (2, 3, 4,...), it is called *extensible*. Such functions have a "yes" in the corresponding column in Table 5.1.

No formal parameters have to be entered when calling extensible functions. In textual languages they are called simply by using actual parameters separated by commas – in graphical representation the parameter names inside the boxes are omitted.

In IEC 61131-3 these properties are **not** applied to user-defined functions, but can also be extended to these functions (and other POU types) as a supplement to the standard, depending on the programming system.

Overloading and extensibility of standard functions are explained with the aid of examples in the next two sections.

Overloaded functions

Overloaded functions can be applied for processing several data types using only one function name.

An overloaded function does **not** always support **every** data type of a generic data type, as explained in Chapter 3.

For example, if a PLC programming system recognises the integer data types INT, DINT and SINT, only these three data types will be accepted for an overloaded function ADD which supports the generic data type ANY_INT.

If a standard function is not overloaded, but restricted to a certain elementary data type, an underline character and the relevant data type must be added to its name: e.g. ADD_SINT is an addition function restricted to data type SINT. Such functions are called **typed**. Overloaded functions can also be referred to as **type-independent**.

This is illustrated in Example 5.1 using integer addition:

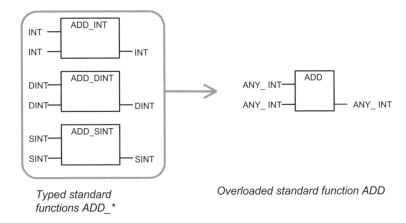

*Typed standard
functions ADD_ **

Overloaded standard function ADD

Example 5.1. Typed standard functions ADD_INT, ADD_DINT and ADD_SINT for integer addition and the overloaded standard function ADD

When overloaded functions are used, the programming system automatically chooses the appropriate typed function. For example, if the ADD function shown in Example 5.1 is called with actual parameters of data type DINT, the ADD_DINT function will automatically be selected and called (invisibly to the user).

All derived data types made available by the PLC system are supported.

When calling standard functions, each overloaded input and, in some cases, the function return value, must be of the same data type, i.e. it is not permissible to use variables of different types as actual parameters at the same time.

If the inputs are of different data types, the PLC programmer must use explicit type conversion functions for the corresponding inputs and function return value respectively, as shown in Example 5.2 for ADD_DINT and ADD_INT. In such cases, instead of the overloaded function ADD its typed variant (e.g. ADD_DINT) should be used.

```
VAR
  Var_Integer        :    INT;
  Var_ShortInteger   :    SINT;
  Var_FloatingPoint  :    REAL;
  Var_DoubleWord     :    DWORD;
END_VAR
```

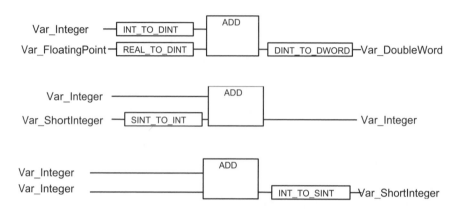

Example 5.2. Calls of the overloaded standard function ADD with type conversion functions to ensure correct input data types. In the top case the programming system replaces ADD with the typed standard function ADD_DINT. In the other two cases the function ADD_INT is used.

Extensible functions

Extensible standard functions can have a variable number of inputs, between two and an upper limit imposed by the PLC system. In graphical representation, the height of their boxes depends on the number of inputs.

Extending the number of inputs of a standard function serves the same purpose as using cascaded calls to the same function, in both the textual and the graphical programming languages of IEC 61131-3. Especially in the graphical languages LD and FBD, the amount of space required to write the function can be greatly reduced.

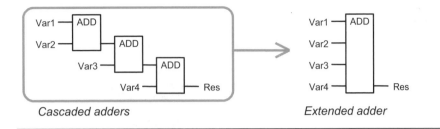

Cascaded adders	Extended adder

Instruction List (IL)

```
LD      Var1
ADD     Var2
ADD     Var3
ADD     Var4
ST      Res
```

Can be replaced by:

```
LD      Var1
ADD     Var2, Var3, Var4
ST      Res
```

Structured Text (ST)

```
Res := Var1 + Var2;
Res := Res  + Var3;
Res := Res  + Var4;
```

Can be replaced by:

```
Res := Var1 + Var2 + Var3 + Var4;
```

Example 5.3. Cascaded functions as an alternative representation for an extensible function showing addition in graphical and textual (IL and ST) representation

In Example 5.3 the triple call of the standard function ADD is replaced by a single call with extended inputs. Simplifications also result for the textual versions in IL and ST.

5.1.2 Examples

In this section the calling interfaces of the standard functions are shown in examples. The subject of calling functions has already been discussed in detail in Chapter 2.

At least one example has been selected from each function group in Table 5.1. The examples are given in the textual languages IL and ST and graphical representations LD and FBD. In IL and ST the names of the formal parameters are not specified explicitly in the function calls.

For these examples the PROGRAM ProgFrameFUN in Example 5.4 is used as the basis for the common declaration part for the required variables.

```
TYPE                                     (* enumeration type for colours *)
   COLOURS :    ( lRed, lYellow, lGreen,      (* light *)
                  Red, Yellow, Green,         (* normal *)
                  dRed, dYellow, dGreen);     (* dark *)
END_TYPE

PROGRAM ProgFrameFUN          (* common declaration part for std. FUNs *)
   VAR                                        (* local data *)
      RPM         : REAL:= 10.5;              (* revs *)
      RPM1        : REAL;                     (* revs 1 *)
      RPM2        : REAL := 46.8895504;       (* revs 2 *)
      Level       : UINT := 1;                (* revs level *)
      Status      : BYTE := 2#10101111;       (* status *)
      Result      : BYTE;                     (* intermediate result *)
      Mask        : BYTE := 2#11110000;       (* bit mask *)
      PLCstand    : STRING [11]:= 'IEC 61131-5';  (* character string *)
      AT %IB2     : SINT;                     (* for MUX selection *)
      AT %QX3.0   : BOOL;                     (* output bit *)
      DateTime    : DT := dt#1994-12-23-01:02:03;  (* date and time *)
      VTime       : TIME := t#04h57m57s;      (* time *)
      TraffLight  : COLOURS;                  (* traffic light *)
      ColScale1   : COLOURS:= lYellow;        (* scale of colours 1 *)
      ColScale2   : COLOURS:= Yellow;         (* scale of colours 2 *)
      ColScale3   : COLOURS:= dYellow;        (* scale of colours 3 *)
      Scale       : INT := 2;                 (* selection scale of colours *)
   END_VAR
   ...                  (* program body with examples *)
END_PROGRAM
```

Example 5.4. Common declarations for the examples explaining the usage of the standard functions

Type conversion functions

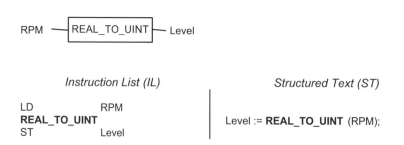

	Instruction List (IL)	Structured Text (ST)

```
LD          RPM
REAL_TO_UINT
ST          Level
```

Level := **REAL_TO_UINT** (RPM);

Example 5.5. Converting from REAL to UINT

This example shows type conversion of the REAL value RPM (floating point) to the unsigned integer (UINT) value Level.

The variable Level has the value 10 after executing the function, as it is rounded down from 10.5.

Numerical functions

	Instruction List (IL)	*Structured Text (ST)*
LD	RPM	
LN		RPM1 := **LN** (RPM);
ST	RPM1	

Example 5.6. Natural logarithm

This example shows the calculation of a natural logarithm. Variable RPM1 has the value 2.3513 after execution.

Arithmetic functions

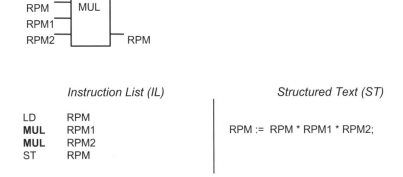

	Instruction List (IL)	*Structured Text (ST)*
LD	RPM	
MUL	RPM1	RPM := RPM * RPM1 * RPM2;
MUL	RPM2	
ST	RPM	

Example 5.7. Multiplication. Instead of using MUL twice in IL, the shortened form: MUL RPM1, RPM2 can also be used.

This example uses the overloaded multiplication function. Because its inputs are of type REAL it is mapped to the typed function MUL_REAL. The variable RPM has the value 1157,625 after execution of the function.

In graphical representation, the multiplication sign " * " may also be used instead of the keyword MUL. This is shown here only for ST.

Bit-shift functions

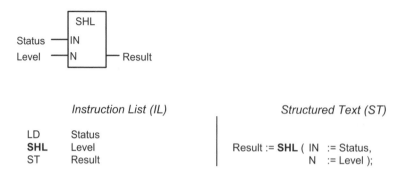

	Instruction List (IL)	*Structured Text (ST)*

LD   Status
SHL   Level   Result := **SHL** (IN   := Status,
ST   Result   N   := Level);

Example 5.8. Bit-shift left

In Example 5.8 the shift function SHL is used to shift the value of the variable Status to the left by the number of bit positions specified by the value Level.

After executing the shifting function, Result has the value 2#01011110, i.e. when shifting left, a "0" is inserted from the right.

Bitwise Boolean functions

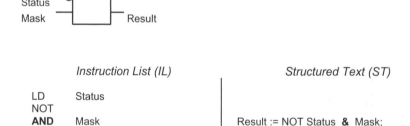

	Instruction List (IL)	*Structured Text (ST)*

LD   Status
NOT
AND   Mask   Result := NOT Status **&** Mask;
ST   Result

Example 5.9. AND operation

As the logical AND is an extensible function, the input parameter names need not be given (similar to MUL). AND is also an overloaded function, its inputs and function value here are of type BYTE. The programming system therefore automatically uses the typed function AND_BYTE.

Instead of the IL instructions LD and NOT in Example 5.9 the Boolean operator LDN (load negated) could be used. This inversion is graphically represented by a function input with a small circle.

In graphical representation, the normal AND symbol "&" may also be used instead of the keyword AND. This is shown here only for ST.

In Example 5.9 AND is used to extract certain bits from the value Status with the aid of a bit mask.

The output Result has the value 2#01010000 after executing the function, i.e. the lower four bits have been reset by the mask.

Selection functions

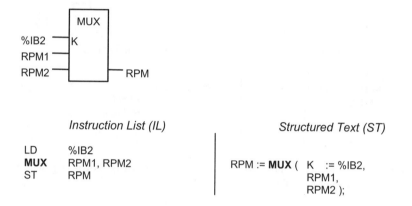

Instruction List (IL)	Structured Text (ST)
LD %IB2	
MUX RPM1, RPM2	RPM := **MUX** (K := %IB2,
ST RPM	RPM1,
	RPM2);

Example 5.10. Multiplexer

The multiplexer MUX has an integer input K and overloaded inputs of the same data type as the function value. The input parameter names, with the exception of K, can therefore be left out. In Example 5.10 K is of data type SINT (integer Byte with sign).

If the input byte %IB2 has the value "1", RPM is assigned the value RPM2 after execution, if it is "0", it is assigned the value RPM1.

If the value of input K is less than 0 or greater than the number of the remaining inputs, the programming system or the run-time system will report an error (see also Appendix E).

Comparison functions

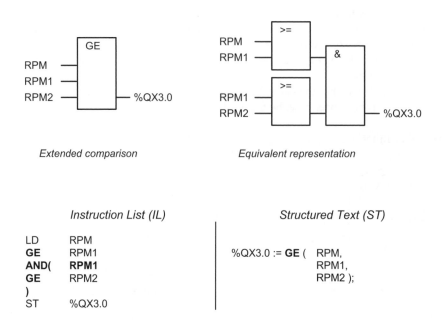

Extended comparison   *Equivalent representation*

Instruction List (IL)	*Structured Text (ST)*

LD   RPM
GE   RPM1
AND(**RPM1**
GE   RPM2
)
ST   %QX3.0

%QX3.0 := **GE** (RPM,
                    RPM1,
                    RPM2);

Example 5.11. An extended comparison. An equivalent solution is shown in graphical representation on the right-hand side.

Comparison functions use overloaded inputs, their output Q is Boolean. They represent a kind of "connecting link" between numerical/arithmetic calculations and logical/Boolean operations.

In the graphical representation of Example 5.11 the comparison function extended by one additional input is also shown as an equivalent call of three functions. Here the key words GE and AND are replaced by the symbols ">=" and "&".

Character string functions

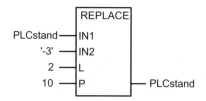

Instruction List (IL)	*Structured Text (ST)*

LD   PLCstand
REPLACE   '-3', 2, 10
ST   PLCstand
	PLCstand:= **REPLACE** (
	       IN2 := '-3',
	       IN1 := PLCstand,
	       P := 10,
	       L := 2);

Example 5.12. Example of "REPLACE" in IL and ST

The function REPLACE has no overloaded inputs. When calling this function, both graphically and in ST, each input parameter must be entered with its name.

Example 5.12 shows, that these inputs can then be entered in any order (see ST example). On the other hand, the order is fixed if there are no input parameter names (see IL example).

The character string PLCstand has the STRING value 'IEC 61131-3' after execution.

Functions for time data types.

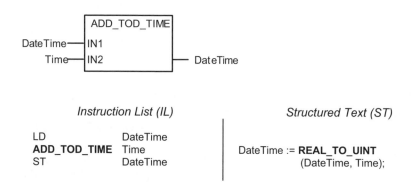

Instruction List (IL)	*Structured Text (ST)*
LD DateTime **ADD_TOD_TIME** Time ST DateTime	DateTime := **REAL_TO_UINT** (DateTime, Time);

Example 5.13. An example of "ADD Time" in IL and ST. The standard now advises against using the "+" sign for time operations.

This time addition function (and also the corresponding subtraction) can be regarded as a continuation of overloaded addition — referring to mixed arguments: TIME, TIME_OF_DAY (TOD) and DATE_AND_TIME (DT).

The variable DateTime has the value DT#1994-12-23-06:00:00 after executing the function.

The addition and subtraction of time is not symmetrical. For subtraction, as opposed to addition, there are three additional operations for input data types DATE, TOD and DT. These operations are not available for addition, as it does not make much sense to add, for example, 10th October to 12th September.

In addition, it is not possible to add a TIME to a DATE, whereas this is possible for TIME, TOD and DT. In order to make this possible with DATE, the input must first be converted to DT and then added. Possible programming errors in time calculations can thus be avoided.

Functions for enumerated data types

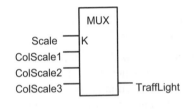

	Instruction List (IL)		*Structured Text (ST)*
LD	Scale		
MUX	ColScale1,	TraffLight := **MUX** (K := Scale,	
	ColScale2,	ColScale1,	
	ColScale3	ColScale2,	
ST	TraffLight	ColScale3);	

Example 5.14. An example of MUX with enumeration

IEC 61131-3 defines functions for the data type enumeration, one of which, the selection function MUX, is shown in Example 5.14.

A variable of data type enumeration (type declaration COLOURS) is selected using the INT variable Scale.

After executing MUX the variable TraffLight will have the values "lYellow", "Yellow" and "dYellow" from the colour scale (light, normal and dark) when the variable Scale has the values 0, 1 and 2 respectively.

5.2 Standard Function Blocks

IEC 61131-3 defines several standard function blocks covering the most important PLC functions (with retentive behaviour).

IEC 61131-3 defines the following five groups of standard FBs:

1) Bistable elements (= flip-flops)
2) Edge detection
3) Counters
4) Timers
5) Communication function blocks.

Table 5.3 gives a concise list of all the standard FBs available in these groups. The table structure is very similar to the one for standard functions in Table 5.1. The communication FBs are defined in part 5 of IEC 61131 and not dealt with in this book.

Instead of the data types of the input and output variables, their names are listed here. These names, together with their corresponding elementary data types, can be found in Table 5.4.

Name of std. FB with input parameter names		Names of output parameters	Short description
Bistable elements			
SR	(S1, R,	Q1)	Set dominant
RS	(S, R1,	Q1)	Reset dominant
Edge detection			
R_TRIG {->}	(CLK,	Q)	Rising edge detection
F_TRIG {-<}	(CLK,	Q)	Falling edge detection
Counters			
CTU	(CU, R, PV,	Q, CV)	Up counter
CTD	(CD, LD, PV,	Q, CV)	Down counter
CTUD	(CU, CD, R, LD, PV,	QU, QD, CV)	Up/down counter
Timers			
TP	(IN, PT,	Q, ET)	Pulse
TON {T---0}	(IN, PT,	Q, ET)	On-delay
TOF {0---T}	(IN, PT,	Q, ET)	Off-delay
Communication			*See IEC 61131-5*

Table 5.3. List of standard function blocks

Inputs / Outputs	Meaning	Data type
R	Reset input	BOOL
S	Set input	BOOL
R1	Reset dominant	BOOL
S1	Set dominant	BOOL
Q	Output (standard)	BOOL
Q1	Output (flip-flops only)	BOOL
CLK	Clock	BOOL
CU	Input for counting up	R_EDGE
CD	Input for counting down	R_EDGE
LD	Load (counter) value	INT
PV	Pre-set (counter) value	INT
QD	Output (down counter)	BOOL
QU	Output (up counter)	BOOL
CV	Current (counter) value	INT
IN	Input (timer)	BOOL
PT	Pre-set time value	TIME
ET	End time output	TIME
PDT	Pre-set date and time value	DT
CDT	Current date and time	DT

Table 5.4. Abbreviations and meanings of the input and output variables in Table 5.3

The counter inputs CU and CD are of data type BOOL and have an additional attribute R_EDGE, i.e. a rising edge has to be recognised in order to count up or down.

The output values of each standard FB are zero when the FB is called for the first time.

The input parameter names of the standard FBs are keywords. In IL they can be applied as operators to FB instances, as described in Section 4.1.4.

The input parameters R and S have a second meaning in IL. There they are also the operators used to set and reset Boolean variables. This can cause difficulties that need to be solved when implementing programming systems.

The concept of overloading inputs and outputs, as described in Section 5.1.1 for standard functions, applies also to standard function blocks.

5.2.1 Examples

In this section, examples are given to illustrate the calling interfaces of standard function blocks in the same way as for the standard functions. The subject of FB calls has already been discussed in detail in Chapter 2.

At least one example is given for each function block group in Table 5.3. Both the textual languages IL and ST and the graphical representations LD and FBD are used.

In IL and ST the FB input parameter names are given explicitly in order to make the use of FB instances as clear as possible.

In the case of IL, the version of the function block call (see Section 4.1.4) that treats input parameters and return values as structure elements of the FB instance is used.

For the following examples, the PROGRAM ProgFrameFB in Example 5.15 is used as the basis for the common declaration part for the required variables and FB instances.

```
PROGRAM ProgFrameFB                      (* common declaration part for std. FBs *)

  VAR_GLOBAL RETAIN                      (* global, battery-backed data *)
    TimePeriod      :   TIME := t#63ms;  (* 63 milliseconds as initial value *)
  END_VAR

  VAR                                    (* local data *)
    FlipFlop        :   RS;              (* flag *)
    Button          :   R_TRIG;          (* edge detection button *)
    Counter_UD      :   CTUD;            (* counter up/down *)
    V_pulse         :   TP;              (* extended pulse *)
    Pulse           :   BOOL;            (* pulse flag*)
    EmOff           :   BOOL;            (* emergency off flag*)
    AT %IX1.4       :   BOOL;            (* emergency off *)
    AT %IX2.0       :   BOOL;            (* count up *)
    AT %IX2.1       :   BOOL;            (* load counter *)
    AT %IX2.2       :   BOOL;            (* start time *)
    AT %IX3.0       :   BOOL;            (* count down *)
    AT %IW5         :   INT;             (* count limit *)
    AT %MX3.2       :   BOOL;            (* flag *)
    AT %QX3.2       :   BOOL;            (* output *)
    MaxReached      :   BOOL;            (* counter at max. limit *)
    MinReached      :   BOOL;            (* counter at min. limit *)
    CounterValue  AT %MW2 : INT;         (* current counter value *)
    TimerValue      :   TIME;            (* current timer value *)
    DateAndTime     :   DT;              (* current date and time *)
  END_VAR
    ...                 (* program body for following examples *)

END_PROGRAM
```

Example 5.15. Common declarations for the examples on the usage of the standard function blocks

These declarations contain:

- FB instances (from FlipFlop to V_pulse)
- Directly represented variables (from %IX1.4 to %QX3.2)
- Symbolic variable (CounterValue)
- General variables (others).

The variables declared in the VAR section are declared as local variables and those declared in the VAR_GLOBAL RETAIN section are declared as battery-backed global variables.

Bistable element (flip-flop)

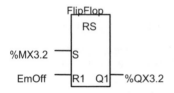

	Instruction List (IL)		*Structured Text (ST)*

```
LD      %MX3.2
ST      FlipFlop.S
LD      EmOff
ST      FlipFlop.R1          FlipFlop (  S := %MX3.2,
CAL     FlipFlop                         R1 := EmOff);
LD      FlipFlop.Q1
ST      %QX3.2
```

Example 5.16. Bistable element (flip-flop)

Example 5.16 shows how to use a flip-flop to store binary status information, in this case the value of flag %MX3.2.

The input R1 "dominantly" resets the output Q1, i.e. if both inputs are set to "1" the output remains "0".

Edge detection

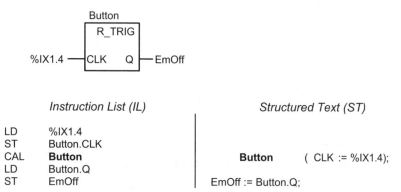

	Instruction List (IL)		Structured Text (ST)
LD	%IX1.4		
ST	Button.CLK		
CAL	**Button**	**Button**	(CLK := %IX1.4);
LD	Button.Q		
ST	EmOff	EmOff := Button.Q;	

Example 5.17. Rising edge detection with R_TRIG

FB instance Button of FB type R_TRIG in Example 5.17 evaluates the signal of an I/O bit and produces a "1" at Q when there is a rising edge (0→1 transition). To do this FB Button uses an internal edge detection flag that stores the "old" value of CLK in order to compare it with the current value.

This information is stored for one program cycle (until the next call) and can be processed by other program parts even if %IX1.4 has already returned to "0" again. At the next call in the following cycle, the Button flag will again be reset.

This means that for directly represented variables FB Button can only detect edges that occur at intervals of at least one program cycle.

IEC 61131-3 provides FBs R_TRIG and F_TRIG not only for immediate usage as shown in Example 5.17. These FBs are also implicitly used for edge detection to implement the variable attributes R_EDGE and F_EDGE (see Chapter 3).

Example 5.18 shows variable declaration using an edge-triggered input (bold text) within the declaration part of FB ExEdge.

```
FUNCTION_BLOCK ExEdge
VAR_INPUT
  Edge        :   BOOL R_EDGE;          (* edge-triggered *)
END_VAR
VAR_OUTPUT
  Flag        :   BOOL;
END_VAR
...
LD    Edge;                             (* access to edge flag *)
ST    Flag;
...
END_FUNCTION_BLOCK
```

Example 5.18. A declaration with R_EDGE for edge detection and usage in IL

To make the use of edge-triggered variables clearer, Example 5.19 shows how additional instructions which implement edge detection are added to Example 5.18. This is done — invisibly to the user — by the programming system.

```
FUNCTION_BLOCK  ExEdge
VAR_INPUT
  Edge        :   BOOL;          (* edge-triggered *)
END_VAR
VAR_OUTPUT
  Flag        :   BOOL;
END_VAR
VAR
   EdgeDetect:   R_TRIG;         (* FB instance "rising edge" *)
END_VAR
...
CAL    EdgeDetect (CLK := Edge);  (* FB call for edge detection *)
LD      EdgeDetect.Q;             (* load detection result from FB instance *)
ST      Flag;
...
END_FUNCTION_BLOCK
```

Example 5.19. Automatic extension of Example 5.18 using R_TRIG

The declaration of FB EdgeDetect in Example 5.19 is inserted implicitly and invisibly by the programming system. This FB is called with input variable Edge. Its output value EdgeDetect.Q is then used wherever the value Edge is originally accessed.

This example shows why IEC 61131-3 does not allow this kind of edge detection for output variables: these variables could be overwritten at any point within the POU. This would, however, violate the rule that FBs are not allowed to change the outputs of other called FBs! See also Section 2.3.2.

Counter

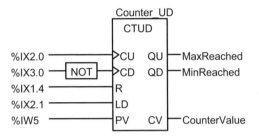

	Instruction List (IL)	Structured Text (ST)

Instruction List (IL)

LD	%IX2.0
ST	Counter_UD.CU
LDN	%IX3.0
ST	Counter_UD.CD
LD	%IX1.4
ST	Counter_UD.R
LD	%IX2.1
ST	Counter_UD.LD
LD	%IW5
ST	Counter_UD.PV
CAL	**Counter_UD**
LD	Counter_UD.QU
ST	MaxReached
LD	Counter_UD.QD
ST	MinReached
LD	Counter_UD.CV
ST	CounterValue

Structured Text (ST)

Counter_UD (CU := %IX2.0
              CD := NOT(%IX3.0),
              R   := %IX1.4,
              LD := %IX2.1,
              PV := %IW5);

MaxReached := Counter_UD.QU;
MinReached := Counter_UD.QD;
CounterValue := Counter_UD.CV;

Example. 5.20. The up/down counter CTUD

In this example, each input of the up/down counter Counter_UD is used. This is, however, not always necessary.

The inputs CU and CD can be activated simultaneously by a rising edge. In this case the current counter value would not change if the minimum or maximum had not already been reached.

Counter_UD in Example 5.20 counts up with each rising edge at %IX2.0 and counts down with each falling edge at %IX3.0. If CU and CD were assigned the same variable or constant, the counter would count up by one increment and immediately afterwards down again by one increment, in response to the signal.

The pre-set counter value at PV is loaded from %IW5 if the load input LD is active when the FB is called. No rising edge is needed in this case.

Timer

Example 5.21 is an example of the usage of timer FBs. It demonstrates clearly how instances of timers maintain their values, especially those of the input parameters, between calls.

In principle, each input variable of a timer (or any FB) can be set immediately before calling. Such run-time parameter changes could be used to allow the same timer to be used to control several process times simultaneously. Such programming is, however, seldom used in practice as it makes the program difficult to read and can easily lead to errors.

It is sufficient to set the pre-set timer value PT for each instance only once, with the first call, and then to re-use it for later invocations. This means that calling the timer primarily serves to **start** the timer with input IN.

The output variables of a timer can be checked at any point in the program, i.e. they need not be evaluated immediately after calling the timer.

The output parameters are set at each call of the timer FB, i.e. they are updated with the current values of the physical timer running in the background. The timer value may therefore become obsolete between two timer calls. Therefore, in order to avoid distorting the desired time control, it must be ensured that the timer FB is called sufficiently frequently in a periodic task, not too long before Q or ET are evaluated.

Output Q shows whether the time has elapsed or not, and output ET shows the time still remaining.

Timers are thus usually called in the following steps:

1) Setting of the timer value
2) Periodic calling with updating
3) Checking of the timer values.

In a PLC program executing periodically, these three steps are often combined in a single call. This simplifies the program and makes the graphical representation easier.

The behaviour of the different timers is shown in more detail in Appendix B.

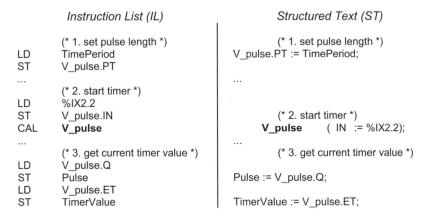

	Instruction List (IL)		Structured Text (ST)
	(* 1. set pulse length *)		(* 1. set pulse length *)
LD	TimePeriod		V_pulse.PT := TimePeriod;
ST	V_pulse.PT		
...			...
	(* 2. start timer *)		
LD	%IX2.2		
ST	V_pulse.IN		(* 2. start timer *)
CAL	**V_pulse**		**V_pulse** (IN := %IX2.2);
...			...
	(* 3. get current timer value *)		(* 3. get current timer value *)
LD	V_pulse.Q		
ST	Pulse		Pulse := V_pulse.Q;
LD	V_pulse.ET		
ST	TimerValue		TimerValue := V_pulse.ET;

Example 5.21. Creating pulses using the timer TP

Example 5.21 shows the three steps required when using the instance V_pulse:

1) The timer value for V_pulse is pre-set to 63 milliseconds.
2) V_pulse is started by input bit 2.2.
3) V_pulse is evaluated by checking Q and ET.

Note:
The timer FB RTC (real-time clock) is no longer included in IEC 61131-3.

6 State-of-the-Art PLC Configuration

IEC 61131-3 takes advantage of recent advances in technology by incorporating modern concepts that allow the modelling of PLC projects consisting of more than just single-processor applications.

The software model of IEC 61131-3 allows for practice-oriented structuring (modularization) of applications into units (POUs). This eases maintenance and documentation, and improves the diagnostics facilities for PLC programs.

A uniform software architecture is essential for the portability of applications. The resources of PLCs are given explicit run-time properties, thus building a platform for hardware-independent programs.

In the traditional method of structuring PLC projects (see Figure 2.5), applications are modularised into blocks, and certain types of blocks (e.g. organisation blocks) have implicit run-time properties. IEC 61131-3 provides more sophisticated and standardised means of accomplishing this.

This chapter explains the *configuration elements* of IEC 61131-3, which are an important means of structuring applications and defining the interaction of POUs. Configuration elements describe the run-time properties of programs, communication paths and the assignment to PLC hardware.

IEC 61131-3 configuration elements support the use of today's sophisticated operating systems for PLCs. The CPU of a typical PLC today can run multiple programs at the same time (multitasking).

6.1 Structuring Projects with Configuration Elements

The preceding chapters have discussed the programming and usage of POUs. This section gives an overview of the modelling and structuring of PLC applications at a higher level.

K.-H. John, M. Tiegelkamp, *IEC 61131-3: Programming Industrial Automation Systems*, 2nd ed., DOI 10.1007/978-3-642-12015-2_6,
© Springer-Verlag Berlin Heidelberg 2010

To do this, Figure 2.7 (POU calling hierarchy) needs to be seen within the context of the PLC program as a whole:

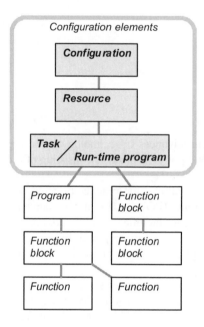

Figure 6.1. Overall structure of PLC programs according to IEC 61131-3, including POUs and configuration elements

As shown in Figure 6.1, the *configuration elements* configuration, resource and task with run-time program are hierarchically located above the POU level.

While POUs make up the calling hierarchy, configuration elements assign properties to POUs:

-   PROGRAMs and FBs are assigned run-time properties
-   Communication relationships are defined between configurations
-   Program variables are assigned to PLC hardware addresses.

First, we will describe the structure and meaning of the configuration elements themselves.

6.2 Elements of a Real-World PLC Configuration

Configuration elements match the elements found in real-world PLC systems:

- *Configuration*:   A PLC system, e.g. a controller in a rack with multiple (interconnected) CPUs, controlling a cell of machines
- *Resource*:   One CPU in a PLC, possibly capable of multitasking
- *Task*:   Run-time properties for programs and function blocks ("type" of PLC program)
- *Run-time program:*   Unit consisting of a PROGRAM or FUNCTION_BLOCK and the TASK associated with it

The main programs of a CPU are made up of POUs of type PROGRAM. Larger applications tend to be structured in Sequential Function Chart, controlling the execution of the other POUs.

Main programs and function blocks are assigned run-time properties, like "cyclic execution" or "priority level", as indicated in Figure 6.2.

The term "run-time program" denotes the unit consisting of all the necessary POUs *and* the TASK, i.e. a program together with its run-time properties. A run-time program is therefore a self-contained unit capable of running independently in a CPU.

Figure 6.2 shows the relation between configuration elements and the components of real-world PLC systems (see also Figure 2.4):

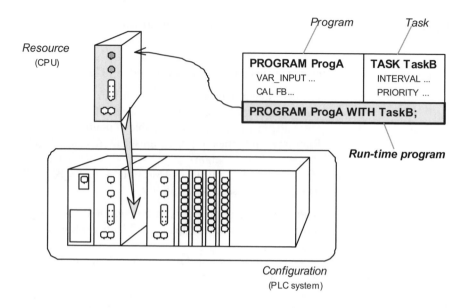

Figure 6.2. A real-world configuration. ProgA and TaskB are linked to form a run-time program and assigned to a CPU resource in a PLC system

The actual assignment of configuration elements to the elements of a PLC system will depend on the hardware architecture.

Using configuration elements, all tasks can be assigned to *one* CPU which will execute them simultaneously, or they can be assigned to different CPUs.

Whether a RESOURCE is to be regarded as one CPU or a group of CPUs contained in one rack therefore depends on the concrete PLC hardware architecture.

For small PLC systems, all configuration can be done in one POU of type PROGRAM: programs can declare global variables and access paths, and directly represented variables. The definition of run-time properties, CPU assignment, etc. can be performed implicitly by features of the programming system or PLC. This corresponds to the traditional approach for programming PLCs.

This capability of PROGRAM POUs facilitates gradual migration from existing applications to IEC 61131-3-compliant programs.

6.3 Configuration Elements

We will first give an overview of the configuration elements, and then explain them in more detail. We will refer to the example in Section 6.4.

6.3.1 Definitions

The functions of the configuration elements are as follows:

Configuration Element	Description
Configuration	- Definition of global variables (valid within this configuration) - Combination of all resources of a PLC system - Definition of access paths between configurations - Declaration of directly represented (global) variables
Resource	- Definition of global variables (valid within this resource) - Assignment of tasks and programs to a resource - Invocation of run-time programs with input and output parameters - Declaration of directly represented (global) variables
Task	- Definition of run-time properties
Run-time program	- Assignment of run-time properties to a PROGRAM or FUNCTION_BLOCK

Table 6.1. Definition of configuration elements. Directly represented global variables and access paths can also be defined within a PROGRAM.

Declaration of directly represented variables maps the entire configuration to the hardware addresses of the PLC. These declarations can be made with VAR_GLOBAL at the configuration, resource or PROGRAM level. POUs can access these via VAR_EXTERNAL declarations.

When put together, the declarations of directly represented variables for all POUs make up the allocation table of a PLC application. Rewiring, i.e. re-assigning symbolic addresses to absolute PLC addresses, can be carried out by simply modifying this list.

Configuration elements are typically declared in textual form. The standard provides a definition for a graphical representation of a TASK, but the graphical representation of all other configuration elements is left to the programming system and is therefore implementation-dependent.

6.3.2 The CONFIGURATION

IEC 61131-3 uses the Configuration (CONFIGURATION) to group together all the resources (RESOURCE) of a PLC system and provide them with means for data exchange. A configuration consists of the elements shown in Figure 6.3.

CONFIGURATION *Configuration name*

Type definitions Global declarations

RESOURCE declarations

ACCESS declarations

END_CONFIGURATION

Figure 6.3. Structure of a CONFIGURATION declaration

Within a configuration, type definitions with global validity for the entire PLC project can be made. This is not possible in other configuration elements.

Communication between configurations takes place via access paths defined with VAR_ACCESS. Variables defined with VAR_GLOBAL are valid only within one configuration, and are accessible to all resources, programs and function blocks of that configuration. VAR_EXTERNAL cannot be used at the configuration level.

Communication blocks for communication between configurations are defined in part 5 of IEC 61131 (see also Section 6.5).

```
CONFIGURATION PLC_Cell1
  VAR_GLOBAL ... END_VAR
  RESOURCE CPU_Conveyor ON CPU_001 ... END_RESOURCE
  RESOURCE CPU_Hammer ON CPU_002... END_RESOURCE
  VAR_ACCESS ... END_VAR
END_CONFIGURATION
```

Example 6.1. Elements of the CONFIGURATION in Example 6.6

Example 6.1 shows part of the declaration of a configuration named PLC_Cell1.

It contains a section with global variables, which are not visible to other configurations, but only accessible from resources CPU_Conveyor and CPU_Hammer and all POUs executing on them.

VAR_ACCESS is used for exchanging data between the resources of the same configuration or different configurations (access paths)

Configurations and resources do not contain instructions like POUs, but solely define the relations between their elements.

6.3.3 The RESOURCE

A Resource is defined in order to assign TASKs to the physical resources of a PLC system. A resource consists of the elements shown in Figure 6.4.

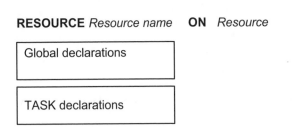

RESOURCE *Resource name*   **ON**   *Resource*

Global declarations

TASK declarations

END_RESOURCE

Figure 6.4. Structure of a resource declaration

The resource name assigns a symbolic name to a CPU in a PLC. The types and numbers of the resources in a PLC system (individual CPU designations) are provided by the programming system and checked to ensure that they are used correctly.

Global variables, which are permissible at resource level, can be used for managing the data that are restricted to one CPU.

```
RESOURCE CPU_Conveyor ON CPU_001
  TASK ...
  PROGRAM ... WITH ...
END_RESOURCE
RESOURCE CPU_Hammer ON CPU_002
  VAR_GLOBAL ... END_VAR
  TASK ...
  PROGRAM ... WITH ...
END_RESOURCE
```

Example 6.2. Elements of the resources in Example 6.6 (Detail from Example 6.1)

Example 6.2 shows part of the declaration of two resources. The global data declared for resource CPU_002 cannot be accessed from resource CPU_001.

The keyword PROGRAM has a different meaning within a resource definition than it has at the beginning of a POU of type PROGRAM!

Within a resource declaration, the keywords PROGRAM ... WITH are used to link a task to a POU of type PROGRAM.

6.3.4 The TASK with run-time program

The purpose of a TASK definition is to specify the run-time properties of programs and their FBs.

The normal practice with PLC systems hitherto has been to use special types of blocks (e.g. organisation blocks, OBs), with implicit, pre-defined run-time properties. For example, they can be used to implement cyclic execution, or to make use of properties of the PLC system for interrupt handling or error responses.

A TASK definition according to IEC 61131-3 enables these program features to be formulated explicitly and vendor-independently. This makes program documentation and maintenance easier.

Figure 6.5 shows the structure of a textual TASK declaration:

TASK Task_name   (Task properties)

PROGRAM Program_name   **WITH**   Task_name : (PROGRAM interface)

Figure 6.5. Structure of a textual declaration of a run-time program, defining a task and associating the task with a PROGRAM. The task properties give the parameter values of the task, the PROGRAM interface gives the actual parameters for the formal parameters.

The association of a TASK with a PROGRAM defines a *run-time program* with the name Program_name. This is the instance name of the program of which the calling interface is given in the declaration. This interface includes input **and** output parameters of the POU of type PROGRAM and is initialised when starting the resource.

One PROGRAM can be executed in multiple instances (run-time programs) using such declarations.

A task can be defined textually, as shown above, or graphically, as shown in Figure 6.6. The task properties are shown as inputs to a box, but the graphical representation of the association with a program (PROGRAM...WITH...) is not defined by IEC 61131-3.

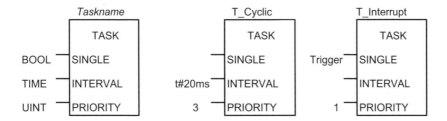

Figure 6.6. Graphical declaration of a task. Left: General form. Centre and right: Two tasks from Example 6.6

The input parameters for tasks shown in Table 6.2 are used for specifying the task properties.

TASK Parameter	Meaning
SINGLE	On a rising edge at this input, programs associated with the TASK will be called and executed once.
INTERVAL	If a time value different from zero is supplied, all programs associated with the TASK will be executed cyclically (periodically). The value supplied is the interval between two invocations. This value can thus be used to set and monitor cycle time. If the input value is zero, the programs will *not* be called.
PRIORITY	This input defines the priority of the associated programs compared to other programs running concurrently (multitasking). The meaning is implementation-dependent (see text below).

Table 6.2. TASK properties as input parameters

The meaning of the PRIORITY input will depend on how concurrency of multiple programs is implemented in the PLC system, and is therefore implementation-dependent. If a task with a priority higher than that of the task currently executing is activated, there are in principle two ways of resolving this conflict between tasks on the same CPU. It depends on the ability of the PLC system to **interrupt** a running task:

1) The task currently executing is interrupted **immediately**, to start execution of the task with higher priority. This is called pre-emptive scheduling.
2) The task currently executing is **not** interrupted, but continues normally until termination. Only then will the task with the highest priority of all waiting tasks be executed. This is called non-pre-emptive scheduling.

Both methods give the task with the highest priority control of the requested resource. If a task with the same priority as the task currently executing is scheduled to execute, it has to wait. If tasks with the same priority are waiting to execute, the one which has been waiting longest will be executed first.

```
TASK T_Quick   (INTERVAL := t#8ms,  PRIORITY := 1);
PROGRAM Motion WITH T_Quick  : ProgA (   RegPar := %MW3,
                                         R_Val => ErrorCode);

TASK T_Interrupt    (SINGLE := Trigger, PRIORITY := 1);
PROGRAM Lube WITH T_Interrupt  : ProgC;
```

Example 6.3. Elements TASK and PROGRAM...WITH... from Example 6.6

In Example 6.3, two tasks T_Quick (cyclic with short cycle time) and T_Interrupt (interrupt task with high priority) are defined.

T_Quick is started every 8 milliseconds. If execution of the program Motion associated with it takes longer than this time (because of interruptions by tasks with higher priority, for example), the PLC system will report a run-time error.

Program Lube has the same priority as Motion, so it has to wait if input Trigger changes to **TRUE**.

Within the assignment of a program to a task, actual parameters can be specified, as shown here for RegPar of ProgA with a directly represented variable. At run time, these parameters are set on each invocation of the run-time program. In contrast to FBs, output parameters can be specified as well as input parameters, and these are also updated at the end of the program. In Example 6.3, this will be done for R_Val of ProgA. Assignments of **output** parameters to variables are specified with "=>" instead of ":=" to distinguish them from **input** parameters.

If a PROGRAM is declared with **no** TASK association, this program will have lowest priority compared to all other programs and will be executed cyclically.

6.3.5 ACCESS declarations

The VAR_ACCESS ... END_VAR language construct can be used to define *access paths* to serve as transparent communication links between configurations.

Access paths are an extension of global variables, which are valid within one configuration only. For access paths, read and write attributes can be specified.

Variables of one configuration are made known to other configurations under symbolic names.

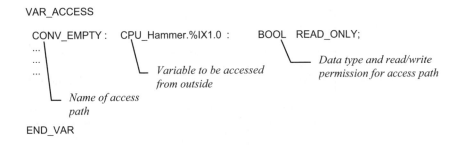

VAR_ACCESS

CONV_EMPTY :   CPU_Hammer.%IX1.0 :   BOOL   READ_ONLY;

END_VAR

Example 6.4. Declaration of an access path

Example 6.4 shows the structure of a declaration of an access path using variable CONV_EMPTY from Example 6.6.

Access paths can be defined for the following types of variables:

- Input and output variables of a PROGRAM
- Global variables
- Directly represented variables.

Access paths publish these variables under a new name beyond the scope of a configuration, so that they can be accessed using communication blocks, for example.

If a variable is a structure or an array, an access path can access only one single member or array element.

By default, access paths allow only read access (READ_ONLY permission). By specifying READ_WRITE, write operations on this variable can be explicitly allowed.

This permission must be specified immediately after the data type of the access variable. The data type of the access variable must be the same as that of the associated variable.

6.4 Configuration Example

Example 6.5 shows an overview of a configuration. This example is declared textually in Examples 6.7 and 6.6. The configuration consists of a PLC system with two CPUs, which are assigned several programs and function blocks as run-time programs. Parts of this example have already been discussed in previous sections of this chapter.

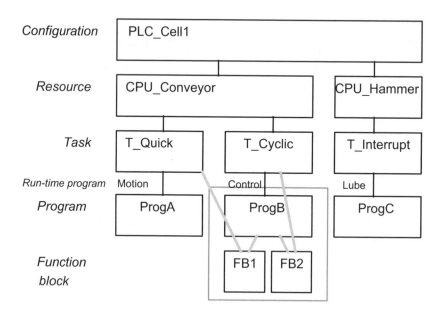

Example 6.5. Example of configuration elements with POUs (overview)

In Example 6.6, PLC_Cell1 physically consists of 2 CPUs. The first CPU can execute two tasks: one fast cyclic task with a short cycle time, and one slower cyclic task. The second CPU executes one task with interrupt property.

```
CONFIGURATION PLC_Cell1

VAR_GLOBAL
  ErrorCode    :   DUINT;
  AT %MW3      :   WORD;
  Start        :   INT;
END_VAR

RESOURCE CPU_Conveyor ON CPU_001
  TASK T_Quick       (INTERVAL := t#8ms,  PRIORITY := 1);
  TASK T_Cyclic      (INTERVAL := t#20ms, PRIORITY := 3);
  PROGRAM Motion WITH T_Quick  : ProgA (RegPar := %MW3);
  PROGRAM Control  WITH T_Cyclic : ProgB (InOut := Start,
                                      R_Val => ErrorCode,
                                      FB1 WITH T_Quick,
                                      FB2 WITH T_Cyclic);
END_RESOURCE

RESOURCE CPU_Hammer ON CPU_002
  VAR_GLOBAL
    Trigger  AT %IX2.5 : BOOL;
  END_VAR
  TASK T_Interrupt     (SINGLE := Trigger, PRIORITY := 1);
  PROGRAM Lube WITH T_Interrupt  : ProgC;
END_RESOURCE

VAR_ACCESS
  RegP          : CPU_Conveyor.Motion.RegPar : WORD      READ_WRITE;
  CONV_EMPTY : CPU_Hammer.%IX1.0           : BOOL      READ_ONLY;
END_VAR

VAR_CONFIG
  CPU_Conveyor.Motion.RegPar.C2      AT QB17 : BYTE;
END_VAR

END_CONFIGURATION
```

Example 6.6. Textual declaration of Example 6.5. Names of variables, programs and FBs are printed in bold type.

Variable C2, the storage location of which was not specified when declared in program ProgA, is assigned a concrete output byte with construct VAR_CONFIG .

```
PROGRAM ProgA          PROGRAM ProgB          PROGRAM ProgC
  VAR_INPUT              VAR_IN_OUT
    RegPar : WORD;         InOut : INT;
  END_VAR               END_VAR
                        VAR_OUTPUT
  VAR                     R_Wert : DUINT;
    C2 AT %Q*: BYTE;     END_VAR
  END_VAR               ...                    ...
                        CAL Inst_FB2           CAL Inst_FB3
                        ...                    ...
END_PROGRAM            END_PROGRAM            END_PROGRAM
```

Example 6.7. Programs for Example 6.6; FB3 is not shown there, FB1 could implement error handling, for example.

In this example, run-time programs Motion, Control and Lube are created by associating programs ProgA, ProgB and ProgC respectively with a task definition.

Program Motion and FB instance FB1 (independent of program Control) are executed on CPU_Conveyor as quick tasks (T_Quick). FB2 of program Control is executed on the same CPU (CPU_001) as a cyclic task (T_Cyclic). Program Control is used here to define the run-time properties of the FB tasks involved. Function block instances associated with tasks in this way are executed **independently** of the program.

With every cyclic invocation of run-time program Control, the input parameter InOut is set to the value of variable Start. After termination of Control, the value of output parameter R_Val is assigned the global variable ErrorCode.

On CPU_Hammer (the second CPU), program Lube is executed as an interrupt-driven task (T_Interrupt). FB3, being part of ProgC, automatically inherits the same run-time behaviour.

In this example, CPU_001 and CPU_002 are not variables, but manufacturer-defined names for the CPUs of PLC_Cell1.

6.5 Communication between Configurations and POUs

This section describes the means of exchanging data between different configurations and within one configuration, using shared data areas.

Such (structured) data areas are used for communication between different program parts, for exchange of data, for synchronisation and to support diagnostics.

It is the aim of IEC 61131-3 to provide a standardised communication model and thus enable the creation of well structured PLC programs, which facilitate commissioning and diagnostics and provide better documentation.

Modularization of applications eases re-use, which helps to reduce the time taken to develop new applications.

IEC 61131-3 defines several ways of exchanging data between different parts of a program:

- Directly represented variables,
- Input and output variables, and the return value, in POU calls,
- Global variables (VAR_GLOBAL, VAR_EXTERNAL),
- Access paths (VAR_ACCESS),
- Communication blocks (IEC 61131-5),
- Call parameters.

The first three methods are for communication **within one** configuration, while access paths and communication blocks are intended for communication **between different** configurations, or with the outside world.

Directly represented variables are not really intended to be used for communication between different parts of an application, but they are included in this list because their use is theoretically possible. Writing to PLC inputs (%I...) is not a suitable method. Outputs (%Q...) should be used to control the process, and not for temporary storage of internal information.

As shown in Table 6.3, these methods can be used at several levels, and the different configuration elements and POUs have different rights of access.

Communication method	CONF	RES	PROG	FB	FUN
Access path	X		X		
Directly represented variable	X	X	X		
Global variable	X	X	X		
External variable			X	X	
Communication block			X	X	
Call parameter		X	X	X	X

Key: CON: CONFIGURATION
        RES: RESOURCE
        PROG: PROGRAM
        FB: FUNCTION_BLOCK
        FUN: FUNCTION

Table 6.3. Communication methods available to configuration elements and POUs.

Access paths are used for exchanging data between configurations, i.e. across the boundaries of one PLC system, and can be used at configuration and program level.

Directly represented variables of a PLC (i.e. %I, %Q and %M) allow limited communication between different parts of an application, as they can be accessed globally on one system. Flags (%M) can be used for synchronising events, for example.

These variables may only be declared at program level or higher (i.e. globally), and function blocks may only access them with an external declaration. This is one important difference between IEC 61131-3 and previous PLC programming practice.

Global variables can be declared for configurations, resources and programs, and can be used at these levels.

Function blocks can access these variables (read and write) with an external declaration, but they cannot declare them themselves. Functions have no access to global or external variables.

External variables can be imported by programs and function blocks if they have been declared globally elsewhere.

Communication blocks are special function blocks used to transfer packets of data from the sender to the recipient. As these FBs are linked to one program, they are local to one configuration and not visible outside.

The definition of such standard communication blocks is contained in Part 5 of IEC 61131 (Communication Services).

Call parameters are used as input and output parameters when calling POUs. They can be used for transferring data into and out of a POU.

As explained in Chapter 2, parameter assignment to input variables and the checking of output variables of a function block can take place independently of its invocation, thereby resulting in characteristics of a communication mechanism which was previously beyond the capabilities of PLC programming.

Resources can pass values to programs when they are associated with tasks, as shown for Motion in Example 6.6. With every invocation, the values are passed as actual parameters or read as output parameters.

7 Innovative PLC Programming Systems

This chapter goes beyond the specifications of IEC 61131-3 and outlines further requirements placed on *programming systems* in the marketplace. These mainly stem from the special conditions to be met in the PLC environment; requirements and solutions are presented based on the IEC 61131-3 programming culture.

7.1 Requirements of Innovative Programming Tools

The performance of a PLC programming system can be judged by three criteria:

- Technological innovation,
- Fulfilment of PLC-specific requirements,
- Cost/benefit ratio.

The fact that IEC 61131-3 no longer makes a strict distinction between PLC, process computer and PC has a major effect on programming. With the aid of various backend compilers, one and the same programming system can be used to generate user code for

- **Compact PLC** (device, mostly in DIN rail format, with integrated I/O connections; proprietary operating system),
- **Modular PLC** (device, mostly 19" PCB, communicating via a multitude of I/O cartridges in the backplane; proprietary operating system),
- **Embedded Controller** (front-end computer with network connection (e.g. Profibus or CAN); proprietary operating system)
- **Embedded PC** (PC hardware-based; mostly Windows CE or (Realtime) Linux)
- **Standard PC** (extended by I/O boards, Windows realtime extension, uninterruptible power supply, …)

All of the diverse combinations can be handled with a single programming system or programming method. As a rule, the programming systems run on Windows XP or Vista and are connected to the target PLC via serial connections or LAN.

K.-H. John, M. Tiegelkamp, *IEC 61131-3: Programming Industrial Automation Systems*, 2nd ed., DOI 10.1007/978-3-642-12015-2_7,
© Springer-Verlag Berlin Heidelberg 2010

This chapter discusses these features and outlines the most important components of a PLC programming system.

Language Editors	Compiler, Linker, System Configurator	Test & Commissioning Communic. Manager	Project Manager	Device Manager	Documentation

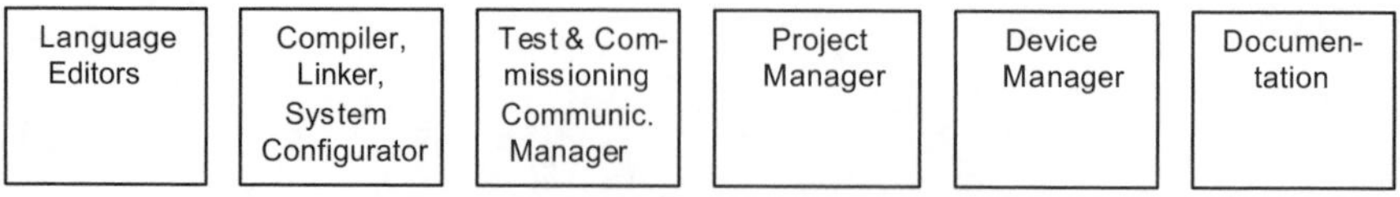

Figure 7.1. Important components of modular PLC programming systems

Supplements to the above-mentioned subpackets (not part of IEC 61131-3) may include:

- Tools for plant project planning,
- Simulation tools,
- General data management systems,
- Parameterisation editors,
- Logic analyser,
- Plant diagnostics,
- BDE system, SCADA system, logging devices,
- CAD connection,
- Network administration.

7.2 Decompilation (Reverse Documentation)

Reverse documentation is a traditional requirement of the PLC market. Ideally, it should be possible to read out a PLC program directly from the controller, in order to modify it on-site at the machine, for example, far away from the office where it was developed. Maintenance personnel want to be able to read, print and modify the PLC program without having the original sources available on a PC.

Decompilation is the ability to retrieve all the information necessary to display and edit a POU directly from the PLC.

IL Code	INTEL 8086 Mnemo-Code (source)

VAR
  V1, V2, Vst: SINT;
END_VAR
 	
LD   V1
AND V2
ST   Vst
	
	MOV AL, V1
	AND AL, V2
	MOV Vst, AL

Example 7.1. Example of decompilable source code. Additional information about the variables is necessary (name, address, type)

Decompilation services can be graded according to the facilities offered:

- No decompilation,
- Decompilation with symbols and comments,
- Decompilation including graphics,
- Sources stored in the PLC.

7.2.1 No decompilation

Most IEC 61131-3 programming systems do not support decompilation. The number of freely selectable symbolic names required to decompile a program has steadily grown and they cannot be stored in the limited memory available on controllers.

It is rare to find PLCs with processor chips specially developed by the manufacturer (e.g. ASIC or bit slice) and using their own specially written machine code. For cost reasons it is more common to use standard processors. It is much more difficult to decompile the machine code from these standard processors back to Instruction List or Structured Text than it is with custom processors.

It is essential to be able to modify programs after commissioning. It is currently state of the art to keep all information related to a project (sources, libraries, interface and configuration information) on a hard disk drive. Ideally, the sources should be in a language-independent form so that they can be displayed in any programming language. Precautions must be taken in the software to ensure that the program on the controller and the program saved on the hard disk are identical before allowing modification.

7.2.2 Decompilation with symbols and comments

The binary code of a program as read out from the PLC will not suffice to create a compilable source. The PLC should provide lists specifying the current wiring (CONFIGURATION). This includes the assignment of symbolic variables to physical addresses, global variables, the mapping of programs to tasks and resources, etc.

Symbolic information (like variable names and jump labels) is typically not contained in the executable code. A symbol table, created during program development and sometimes including comments on declarations and instructions, must be stored in the PLC to enable decompilation of the program directly from the PLC.

7.2.3 Decompilation including graphics

The PLC contains executable code. To be able to display this code graphically on the PC (in Ladder, Function Block Diagram or SFC), the code must either conform to certain syntax rules or it must be augmented by additional information. Using the first method results in shorter programs, but restricts the facilities for graphical representation.

7.2.4 Sources stored in the PLC

The complex architecture of today's IEC 61131-3 programming systems makes it more and more difficult to pack all the information into the binary code. To have all the information needed for decompilation available on the PLC, a simple solution is to store the entire project information in compressed format in a separate slow, low-cost memory within the PLC. From there, this information can easily be transferred to the PC and edited using the programming system in the same way as during program development.

7.3 Language Compatibility

The five languages of IEC 61131-3 have a special relationship with one another.

Sequential Function Chart with its two methods of representation (textual and graphical) is different from the other languages because it is not used for formulating calculation algorithms, but for structuring programs and controlling their execution.

The logic operations and calculations themselves are defined in one of the other languages and invoked from SFC via action blocks, see Examples 4.54 and 4.55 in Section 4.6.6.

Each program consists of blocks (POUs), which invoke each other by means of calls. These blocks are independent of each other and can be written in different languages, even languages not defined by IEC 61131-3, provided the calling conventions of IEC 61131-3 are observed.

IEC 61131-3 does not go beyond defining common calling interfaces between blocks written in different languages. The question is:

> Is it necessary to be able to display code written in one IEC 61131-3 language in another IEC 61131-3 language?

The use of different languages within the same program and the ability to display, print and edit POUs in any IEC 61131-3 language is discussed in the next two sections under the headings **Cross-compilation** and **Language independence**.

7.3.1 Cross-compilation

IEC 61131-3 does not require that a POU developed in one language should be able to be displayed in another language. The argument about the need for this has been going on as long as PLCs have been in existence.

What are the reasons for asking for this feature?

The motivation for cross-compilation
One important reason for wanting to be able to cross-compile parts of a program are the different levels of education and areas of activity of technicians and engineers. Depending on their field, they tend to be trained in different programming languages, making it difficult for them to work together.

In the automotive industry in the US, Ladder Diagram is the preferred language, while the same industry in Europe prefers Instruction List or Structured Text. In the plant construction industry, a functional language like FBD will be preferred. A computer scientist should have no difficulties using Structured Text. So, do we need a different language for every taste, but with editing facilities for all?

Some languages are better suited for certain problems than others. For example, memory management routines are obviously easier to read and write in IL or ST than in Ladder Diagram. A control program for a conveyor is clearer in Ladder Diagram than in ST. SFC is the best choice for a sequential control system.

In many cases, it is not so easy to select the right language. The same section of a program is frequently even needed by different users.

For example, a PLC manufacturer may provide POUs written in IL to support users in handling I/O modules. The user may be a conveyor belt manufacturer, using the PLC to monitor and control limit switches and motors, and preferring to work with Ladder Diagram. So the programmer modifies the code provided in IL to his needs, using Ladder Diagram to do so. The conveyor may then be supplied to a plant construction company where all programs are written in FBD, and the I/O control programs will be required for complete and uniform documentation.

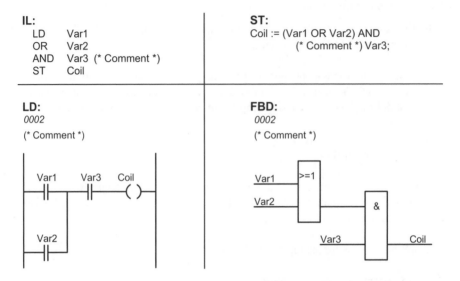

IL:
    LD   Var1
    OR   Var2
    AND   Var3 (* Comment *)
    ST   Coil

ST:
Coil := (Var1 OR Var2) AND
              (* Comment *) Var3;

LD:
0002

(* Comment *)

FBD:
0002

(* Comment *)

Example 7.2. Example of cross-compilation between four different languages of IEC 61131-3

Different approaches in graphical and textual languages.
One difficulty with cross-compilation lies in the different ways of looking at a calculation. LD and FBD have their roots in Boolean or analogue value processing: there is "power flow", or not; values are propagated and calculated in parallel and not clocked. Textual languages, like IL and ST, are procedural, i.e. instructions are executed one after the other.

This becomes obvious when looking at the network evaluation in Section 4.4.4. Example 7.3 gives an example in Ladder Diagram (FBD would be similar).

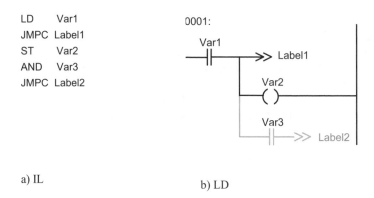

a) IL   b) LD

Example 7.3. Sequential execution of a program section in IL compared to parallel execution of a network in Ladder Diagram. This makes cross-compilation difficult.

According to the evaluation rules given for graphical languages (see Section 4.4.4), Var2 will always be assigned a value. If this network is converted from IL as it stands, Var2 will be assigned a value only if Var1 equals FALSE. Otherwise it would be necessary to rearrange the elements before cross-compiling (all ST instructions before a conditional JMP or CAL). This would change the graphical appearance of the network when cross-compiled to Ladder Diagram.

Looking at the procedural (IL) sequence and the simultaneous evaluation in Example 7.3, the IL sequence converted to Ladder is misleading. When Var1 and Var3 are TRUE, Label1 and Label2 are also TRUE. The IL sequence jumps to Label1; in the Ladder Diagram version both labels are addressed in accordance with the evaluation rules of Ladder, and the next network to be activated is unclear.

This problem of cross-compilation is solved in many of the programming systems that possess this functionality by:

- not allowing further logic operations after an assignment in graphical networks,
- evaluating the code part of a graphical network from top to bottom (Example 4.40) and stopping evaluation when a control flow instruction is being executed.

Differences in languages affect cross-compilation.
Not all of the languages can be cross-compiled to each other. SFC is a language for structuring applications, making use of the other languages, but having a completely different design. We shall therefore only discuss cross-compilation between IL, ST, LD and FBD.

Restrictions in LD/ FBD.
Jumps can be made using labels, but are somewhat contradictory to the concept of "parallel" networks. Some functions, like management of system resources (stack operations), can only be expressed in very complicated, unreadable programs.

Constructs like CASE, FOR, WHILE or REPEAT are not available in these languages and can only be implemented by using standard functions like EQ and complex network arrangements.

Unlike ST, these two languages allow only simple expressions to be used to index arrays.

LD is designed to process Boolean signals (TRUE and FALSE). Other data types, like integer, can be processed with functions and function blocks, but at least one input and one output must be of type BOOL and be connected to the power rail. This can make programs hard to read. For non-Boolean value processing FBD is better suited than LD.

Not all textual elements have a matching representation in the graphical languages. For example, some of the negation modifiers of IL are missing in LD and FBD: JMP, JMPC, RET and RETC are available, but JMPCN and RETCN are not. These can be formulated by the user with additional logic operations or supplementary (non-standard) graphical symbols can be included in the programming system.

Restrictions in IL/ ST.
The notion of a network, as used in the graphical languages LD and FBD, is not known in IL or ST.

In Ladder Diagram, attributes like edge detection (P, N) can be expressed by graphical symbols. This representation of attributes in the code part of a program does not comply with the strict concept of expressing the attributes of variables in the declaration part. There are no matching elements in the textual languages for these graphical symbols.

The use of EN and ENO poses another problem, as no matching element is available in IL or ST. Each user-defined function must evaluate EN and assign ENO a value to be useable in graphical languages. If EN and ENO are used in graphical languages and not used in textual languages, two versions of standard functions are needed (with and without EN/ENO processing), see Section 2.7.1.

Cross-compilation IL / ST.

A high-level language (ST) can be converted more easily and efficiently into a low-level, assembler-like language (IL) than vice versa. In fact, both languages have some features that make cross-compilation into the other difficult. For example:

- ST does not support jumps, which have to be used in IL to implement loops. This makes cross-compilation of some IL programs difficult.
- IL supports only simple variables for array indices and actual parameters for function and function block calls, whereas in ST complex expressions may also be used.

Full cross-compilation only with additional information.

If a system in Ladder Diagram allows multiple coils and/ or conditional instructions with **different** values within a network (corresponding to a one-output to many-input situation in FBD), see Example 7.4, auxiliary variables have to be used to cross-compile to IL. When cross-compiling this code part from IL or ST to Ladder Diagram, care has to be taken to avoid multiple networks being generated in the Ladder Diagram version.

Small modifications of an IL program can, therefore, result in major changes in the Ladder cross-compiled form.

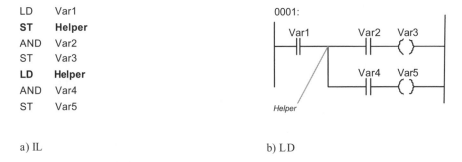

a) IL   b) LD

Example 7.4. If coils or conditional instructions are used in an LD network with different value assignments, direct cross-compilation is not possible. To cross-compile to IL, an auxiliary variable like Helper is necessary.

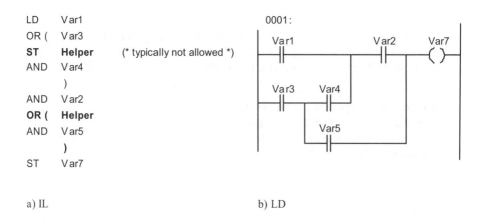

```
LD    Var1
OR (  Var3
ST    Helper    (* typically not allowed *)
AND   Var4
      )
AND   Var2
OR (  Helper
AND   Var5
      )
ST    Var7
```

a) IL   b) LD

Example 7.5. The limits of cross-compilation are reached when it comes to the compilation of complex parenthesised constructs and helper variables into a textual language.

The Ladder network in Example 7.5, which looks perfectly clear in this graphical language, is not directly cross-compilable to IL. An auxiliary variable, like Helper, is necessary to temporarily store the intermediate result between Var3 and Var4, and several parentheses are needed. Storing a variable within a parenthesised expression is not explicitly prohibited in IEC 61131-3, but should be avoided because comparable operators, like conditional jumps, may also not be used within parenthesised expressions.

Some programming systems automatically place graphical elements in the correct (optimum) position, depending on their logical relation. Other systems allow and require users to position the elements themselves. This topographical information needs to be stored in the form of comments in IL or ST, if a graphical representation is to be regenerated from the textual representation later. IL or ST programmers cannot reasonably be expected to insert this topographical information. In most programming systems, the information about the position of the graphical elements is therefore only kept in internal data storage and is lost when the program is cross-compiled to IL or ST. In most cases, there is no way back from the textual languages to the original graphical layout.

Quality criteria for cross-compilation.
It has been explained that full cross-compilation in the theoretical sense cannot be achieved. The quality of a programming system with respect to cross-compilation

depends rather on how well it meets the following conditions:

1) The rules of cross-compilation should be so easy to understand that the programmer can always determine the result.
2) Cross-compilation must not change the semantics of the POU (only local modifications, entities must stay together).
3) Cross-compilation should not affect the run time of the program.
4) Cross-compilation must not introduce side-effects (i.e. not affect other parts of the program).
5) Ambiguities must be resolved (see Example 7.3).

As cross-compilation is not easy to achieve, many programming systems do not implement it.

7.3.2 Language independence

The POU, as an independent entity, is an important item of IEC 61131-3. To invoke a POU in a program, only the external interface of the POU needs to be known, and no knowledge about the code contained within it is required. To design a project top-down, it should therefore be possible to define all the interfaces of the POUs first and fill in the code later.

Once the external interface of a POU has been defined, it is typically made available to the whole project by the programming system. It consists of:

- the function name and function type, plus the name and data type of all VAR_INPUT and - if applicable - VAR_OUTPUT parameters for a FUNCTION,
- the function block name, the name and data type of all VAR_INPUT, VAR_OUTPUT, VAR_IN_OUT parameters and EXTERNAL references for a FUNCTION BLOCK,
- the program name, the name and data type of all VAR_INPUT, VAR_OUTPUT, VAR_IN_OUT and VAR_ACCESS parameters and GLOBAL variables for a PROGRAM.

One POU can invoke another function or function block instance without knowing which language the other POU has been programmed in. This means that the programming system does not have to provide a separate set of standard functions and function blocks for each programming language.

This principle can even be extended to languages outside the scope of IEC 61131-3 as the caller needs no knowledge of the invoked block apart from its external interface. If the external interface and the calling conventions of an IEC 61131-3 programming system and a C compiler, for example, are compatible, it is equally possible to invoke a C subprogram.

IEC 61131-3 expressly allows the use of other programming languages.

7.4 Documentation

Different types of documentation are required to allow for efficient maintenance of applications and to support modern quality standards like ISO 9000:

1) *Cross-Reference List.* A table listing which symbols (variable name, jump label, network title, POU type or instance name etc.) are being used in which POUs.
2) *Program Structure*, giving an overview over the calling hierarchy of POUs. Each POU has a list giving the names of all POUs invoked from it. The Program Structure can be visualised graphically, or textually, with a nesting depth depending on the system.
3) *Allocation List (Wiring List).* A table giving the physical addresses of I/Os and the names of the variables assigned to these addresses.
4) *I/O Map.* A table of all I/O addresses used by the application, sorted by address. The I/O map is helpful in finding free I/O addresses when extending an application and for having the relation between PLC software and PLC hardware documented in a hardware-oriented manner.
5) *Plant Documentation.* A description of the entire plant, typically graphical. Each individual PLC will be only one "black box" in this documentation. The entire plant contains multiple PLCs, machines, output devices etc. The plant documentation is often generated with standard CAD programs, giving the topological grouping and connections between PLCs and other devices.
6) *Program Documentation.* Sources of POUs created with the programming system. When printed, these should closely match in structure and contents the representation on-screen while editing.
7) *Configuration.* The Configuration – as understood by IEC 61131-3 – describes which programs are to be executed on which PLC resources and with what run-time properties.

These types of documentation are neither required nor standardised by IEC 61131-3, but have become popular over the years for documenting PLC programs.

7.4.1 Cross-reference list

The cross-reference list consists of:

- all symbolic names used in a program (or occasionally in an entire project),
- the data type of all variables, the function block type for instance names, the function types for functions plus declaration attributes (like RETAIN, CONSTANT or READ_ONLY) and the variable type (VAR_INPUT, ...),
- the name and type of the POU, including the numbers of all lines in which these variables and/or an instance with this POU type occur,

- the kind of access to this variable at this program location,
- the location of the declaration of this variable.

Some systems support a cross-reference list only for individual POUs, or provide it only for global and external data.

The cross-reference list is helpful in finding program locations referencing variables during debugging.

Different sorting criteria are usually supported. The entries are usually sorted alphabetically by symbolic name. They can also be sorted by symbol type (input, output, function name, instance name,...) or data type (BYTE, INT,...).

Symbolic name	POU name	POU type	Line No.	Access	Data type	Attribute	Var type
Temp_Sensor	SOND.POU	Prog	10	Decl	INT	Retain	GLOBAL
	SOND.POU	Prog	62	Read			GLOBAL
	CONTR.POU	FB	58	Write			EXTERN.
. . .							

Example 7.6. Example of a cross-reference list sorted by symbolic name

7.4.2 Allocation list (wiring list)

The allocation list lists all variables which are assigned to physical I/O addresses of a configuration, plus the access path, if supported.

Sometimes, tools are provided for changing the assignment of symbols to physical addresses (*rewiring*). This often needs to be done when porting an application to another environment (e.g. another PLC with different I/O connections).

Symbolic name	I/O address
Temp_Sensor_1	%IB0
Temp_Sensor_2	%IB2
Temp_Control	%QB4
Temp_Save	%MB1
. . .	

Example 7.7. Example of an allocation list (wiring list)

7.4.3 Comments

Comments are an important part of program documentation. Application sources can be enhanced by descriptive comments at many locations (see Example 7.2):

- In ST and IL, comments can be inserted wherever space characters are allowed,
- IEC 61131-3 does not include any guidelines on comments in the graphical languages. However, network comments, preceding and describing a network, are a valuable aid in documentation.

Some programming systems have menu settings to prevent (accidental) overwriting of the program and allow, for example, changes to be made in the comments only by service personnel.

7.5 Project Manager

The task of the Project Manager is to consistently manage all information related to the implementation of a project. This includes:

- **Source information**
  - The sources of all POUs created, with
    - type definitions, declarations of global data, definition of access paths with VAR_ACCESS,...
    - descriptions of the call interfaces of all POUs in order to check their usage,
  - Version control of all sources,
  - Access restrictions, sometimes with different access levels authenticated by passwords, for:
    - modifying POUs,
    - printing programs,
    - editing libraries.
- **Object information**
  - Compiled sources (intermediate code, object code, executable files),
  - Project creation procedures (call-dependence, creation and modification information for controlling time-dependent compiling operations, for example in MAKE or batch processes),
  - Libraries (standard functions, standard function blocks, manufacturer-defined blocks, communication function blocks, user libraries).

- **Online information**
  - Data for assigning parameters to the project (recipes),
  - Device and configuration information (PLC hardware, I/O modules,...),
  - Additional information for online testing (symbol information, breakpoints,...),
  - Communication information (manufacturer-specific transfer protocols, interfaces).
- **Documentation** (e.g. cross-reference list, allocation list, program structure,...).

The Project Manager administers and archives all this data. Why is a standard file manager (e.g. the Windows Explorer) not sufficient?

POUs are interdependent. Multiple programs within one project can use the same POU, although it only exists as a single file. Function names and function block types have global scope throughout a project (whereas the scope of function block instances is limited to the POU they are defined in, unless they are explicitly declared GLOBAL).

For every invocation of a POU (instance or function) the compiler and editors need interface information about the POU being invoked (types and names of parameters). In order to reduce the overhead of gathering the same information again and again every time it is needed, the Project Manager can store such information and supply it to other parts of the programming system when requested.

Figure 7.2 shows the directory of a hard disk drive (D:), containing sub-directories. In order to visualise the **structure** of a project, it is necessary to evaluate interdependences between calls and environmental factors to enable the relations to be displayed, see Figure 7.3.

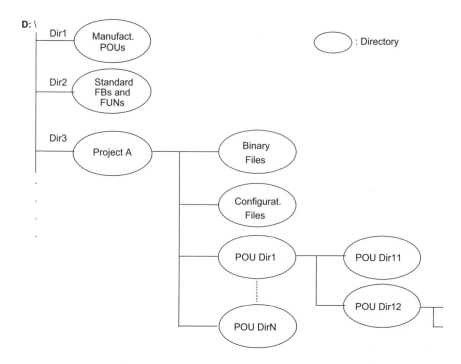

Figure 7.2. Data for sources and libraries stored on disk (shown as drive D:\). The programming system should allow the user to create directories and sub-directories to resemble the project structure as far as supported by the operating system (e.g. Windows).

The programming system should assist the programmer in understanding the structure of the project. One project can contain several configurations (PLCs), and each configuration can contain several resources (CPUs). See Chapter 6 for the description of a configuration. Each resource will have a specification of the hardware associated with it (Resource1 File). Each resource can execute multiple programs (Program1, ...), which are implemented in the form of POUs invoking other POUs. Two distinct instances of one POU, as shown in Figure 7.3, contained in different programs, can be described by the same POU stored in only one file on disk.

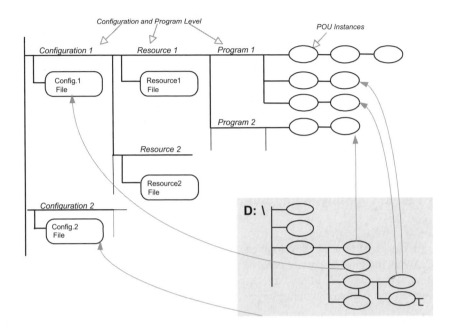

Figure 7.3. The logical structure of a project consists of configurations, resources and a POU calling hierarchy. As this structure in most cases is not identical with the physical file and directory structure, e.g. Figure 7.2, additional visualisation aids are helpful.

Rules can be established for finding information. For example, to compile Program1 for Resource1, all user, system and manufacturer POUs used in the program have to be collected. If one of the POU names called appears more than once on the disk, the system has to take appropriate action (e.g. apply a rule to choose one, or issue an error message); IEC 61131-3 does not discuss this problem.

For efficiency reasons, it is desirable to recompile only a minimum set of sources to create a new version of a resource, configuration or program after a change. For example, if only Program1 has been modified, other programs, or even other resources, that do not need the modified POU need not be recompiled.

A *Project Manager* should therefore support

- registration of newly created files (sources, error logs, interface descriptions),
- import of files from different projects,
- display of all existing POUs,
- renaming and deletion of POUs,
- an information structure that makes the structure of the project (calling hierarchy, dependencies) understandable to the user,

- maintenance of all information (POUs, interfaces, calling hierarchy, binary files, etc.) that the rest of the programming system needs for creating a project.

Some IEC 61131-3 programming systems store data in individual files (e.g. one file per POU) on the hard disk, while others use a database. Either way, project files should only be edited via the Project Manager. To increase efficiency when supplying data in response to requests (e.g. from the linker), some information can be pre-processed, so that it is not necessary to scan the entire file system for each request.

7.6 Test & Commissioning Functions

The first stage of program development is typically performed on a PC, without a PLC. When the most important parts of the program have been completed, work is continued with the PLC in the target environment. The following are typical tasks performed at this commissioning stage:

- Download of the entire project or individual POUs (after modifications) to the PLC,
- Upload of the project from the PLC to the PC,
- Modification of the program in the PLC (either in "RUN" or "STOP" mode),
- Starting and stopping the PLC,
- Display of variable values (status),
- Direct setting of I/Os or variables (*forcing*),
- Deactivation of the physical outputs of the PLC to prevent unsafe plant conditions during tests. Programs are executed and values are assigned to direct variables just as they would be in normal operation. Additional software or hardware ensure that physical outputs are not influenced by values written to the output variables.
- Retrieving PLC system data, communication and network information from the PLC,
- Program execution control (*breakpoint, single step,...*).

These functions are implemented in different ways and to different degrees by individual programming systems. IEC 61131-3 does not stipulate any requirements with respect to these features.

7.6.1 Program transfer

After a POU has been created with the editor and been checked for syntax, PLC-specific code is created. This can take the form of machine code for direct execution by the PLC's processor, or it could consist of instructions to be interpreted by the PLC. The programming system puts this object code together with the object code of other POUs to make a program (link procedure).

All that remains to be done is:

- Connect the executable code with a defined CONFIGURATION (task association),
- Map the logical hardware addresses to the actual physical addresses of the PLC,
- Assign parameters to the PLC's operating system.

These tasks can be performed by a tool called a *System Configurator*, or in some implementations the PLC's operating system can do some of the work.

The entire project now needs to be transferred to the PLC. A tool known as the *Communication Manager* establishes a physical and logical connection to the PLC (e.g. a proprietary protocol via USB interface or a fieldbus protocol). This tool performs some checks, which are partly invisible to the user. For example:

- Has contact been successfully established with the PLC?
- Does the PLC's current state allow transfer of new programs? Or can the PLC be put into the correct state?
- Is the PLC's hardware configuration compatible with the requirements of the program?
- Is the current user authorised to access this PLC?
- Visualisation of the current state of communication.

The program is then downloaded to the PLC, together with all the information needed to run it. The PLC program may now be started (cold restart).

7.6.2 Online modification of a program

If it is necessary to modify blocks whilst the program is running (PLC in "RUN" mode), this can be done in various ways:

- By changing the POU on the PC and compiling the whole program again with the programming system. Everything is then downloaded to the PLC. If this is to be done in "RUN" mode, a second memory area must be available and activated after the download is completed, as the download generally takes too long to suspend execution of the cyclic PLC program.
- By modifying the POU on the PC and downloading only the modified POU to the PLC. This requires a block management function to be available on the PLC, which will accept the new block and replace the old block with the new one after the download has been completed. As the old and new blocks usually differ in size, a "garbage collection" is required from time to time in order to be able to re-use memory space which would otherwise be wasted.

- By replacing only individual networks (SFC, LD, or FBD) or instructions (SFC, ST, and IL). This is only possible if other parts of the POU are not affected. E.g., jump labels in other networks must not move if the PLC operating system does not include jump label management.

7.6.3 Remote control: Starting and stopping the PLC

The PLC hardware typically features a switch to start and stop the PLC. This can be remote-controlled from the programming system.

IEC 61131-3 defines different *start modes* for a PLC (see Section 3.5.3):

1) **Cold Restart**. The PLC starts the program without memorising any variable values. This is the case, for example, after downloading the program to the PLC.
2) **Warm Restart**. Following a power outage, program execution is resumed at the point where it was interrupted (e.g. in the middle of an FB). All variables carrying the RETAIN attribute retain the value they had before the interruption, all other variables are reset to their initial value.
3) **Warm restart at beginning of program**. The values of all RETAIN variables are also retained and all other variables are re-initialised, but the program is restarted at the beginning. This takes place automatically if the interruption time exceeds a parameterised time limit.

As a rule, the user can only initiate cold restarts, whereas warm restarts take place automatically after power recovery.

Additional commissioning features are:

- Stopping of the PLC, either with the current output values or with "safe" output values,
- Deletion of memory areas, to prevent uncontrolled restarts,
- Selection of special operating system modes, e.g. test mode, maximum cycle time, etc.

7.6.4 Variable and program status

The most important test function for debugging and commissioning a PLC program is the monitoring of the status of variables ("flags") and external I/Os. Ideally, values should be displayed in a user-selectable form, as shown in Example 7.8.

Variable	Type	Value
Coil	BOOL	1
Var1	BOOL	1
Var2	BOOL	0
Var3	BOOL	1

a) Variable List

LD	Var1	1
OR	Var2	**0**
AND	Var3	1
ST	Coil	1

Coil := Var1 OR Var2 AND Var3
 1   1   **0**   1

b) IL (Variable Status)

c) ST (Variable Status)

0002:
(* comment *)

d) LD (Variable Status)

0002:
(* comment *)

e) FBD (Variable Status)

LD	Var1	1
OR	Var2	**1**
AND	Var3	1
ST	Coil	1

Coil := Var1 OR Var2 AND Var3
 1
        1
               1
                        1

f) IL (Current Result)

g) ST (Expression)

Example 7.8. Status view of the sample program from Example 7.2 during debugging and commissioning, shown in the different programming languages. The different display modes can be combined. (Example continued on next page)

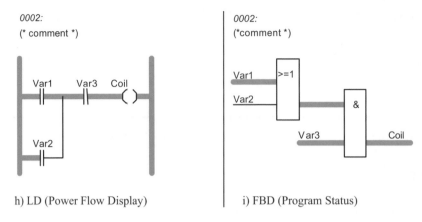

h) LD (Power Flow Display)   i) FBD (Program Status)

Example 7.8. (Continued)

Depending on the implementation in the PLC and/or the programming system, there are different methods of viewing the current data and execution flow:

1) **Variable List**: Display of a list of variables (a). The variables contained in the list are scanned in the PLC and their values continuously updated on the screen. This method is frequently used for monitoring values from different parts of the program. Structured variables can also be displayed showing the values of their individual members.

2) **Variable Status**: All variables of a specific code portion are displayed. In Example 7.8, the values of Var1, Var2, Var3 and Coil are displayed in a separate window (a) or directly within a graphic (b-e).

3) **Program Status** (also **Power Flow**). Instead of displaying the values of individual variables, the result of each operation is displayed (the Current Result, see Section 4.1.2) (f-i). In the graphical languages, this is done by drawing thick or thin lines to represent TRUE and FALSE for Boolean expressions or displaying numerical values beside each line. Intermediate value computation on the PC can serve to display values that are not immediately linked to variables (e.g. operator NOT; OR connection in LD).

The quality of values provided will depend on the functionality, the speed and the synchronisation between the programming system and the PLC. Depending on the operating system and the hardware features available, the values on the PLC can be "collected" at different times:

- Synchronously,
- Asynchronously,
- On change.

Synchronous Status: Values are collected at the end of a program cycle. The values are all generated at the same point in the execution of the program, and are therefore consistent. This is also the time when variables mapped to I/O addresses are written to the outputs (update of the process image). A cyclic program will in most cases execute much faster than values can be retrieved by the programming system, so the values viewed with the programming system are only updated every *n* cycles.

Asynchronous Status: The programming system constantly requests a new set of values for the specified variables (asynchronously to the cycle of the PLC program). The values are collected at the moment when the request is received. Each value is a snapshot of the respective memory location.

Status on change: Values are only collected when they change (e.g. a signal edge). This requires special hardware (address monitoring) within the PLC.

Advanced systems supply additional *data analysis tools*. These allow the behaviour of variable values to be visualised over time (logic analysis) as shown in Figure 7.4.

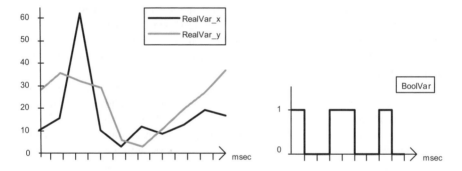

Figure 7.4. Example of data analysis of two REAL variables and one BOOL variable

Visualisation in program graphics quickly reaches its limit when it comes to displaying arrays or complex structures and their values. Separate variable windows are better suited for this purpose.

Values for display include variables as well as parameters of functions and function blocks.

A facility that is not often implemented as yet is the visualisation of the data flow between individual POUs, as shown in Figure 7.5.

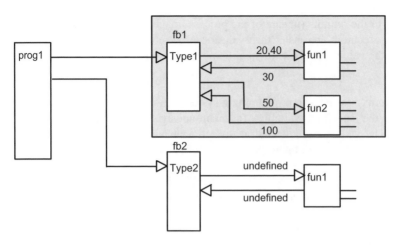

Figure 7.5. Display of the calling hierarchy of the POUs together with the actual parameters during execution of the PLC program.

When requesting variables to be displayed, it is not sufficient to specify the name of the POU in which they occur. In the example shown in Figure 7.5, the name fun1 would not be sufficient to identify which variables to display, as this POU is called twice (Type1 and Type2) and will return different values depending on the call parameters.

In the case of function blocks, it is not sufficient to specify the name of the POU, as the function block type may be instantiated more than once and, as a result, will have different instance data. Additional specification of the instance name is also not unambiguous because a function block type may be instantiated with the same instance name in different POUs so that several instances with the same name will be used.

For unambiguous identification it is therefore necessary to specify the call path and, in the case of functions, even the call location.

Additional difficulties arise from the fact that the local variables of functions only exist during execution of the function, and are otherwise undefined.

7.6.5 Forcing

To test the behaviour of programs and plant it is helpful to be able to force variables to specific values.

To simulate a specific plant condition for the program in order to test certain program parts, some variables are set to fixed values using the programming system, and the PLC is made to use these values instead of the actual values.

			Local Variables
Variable	**Data Type**	**POU Instance**	**Assigned Value**
Var_Loc	BOOL	FB1	1
Var3	INT	FB1	20
...			

Directly represented variables (WORD)

%IW0: **0 0 1 0 0 0 1 0 0 0 0 0 0 0 1 0**   *(Bits 0 to 15)*

%IW1: **0 0 0 0 0 0 0 0 1 1 0 0 0 0 0 0**   *(Bits 16 to 31)*

...

		Symbolic Variables (Integer)
Start Address	**Variable Name**	**Assigned Value**
%QW20:	LimitValue	11.233
...		

Example 7.9. Example of the forcing of variables and I/O addresses (Boolean and integer values). When specifying the name of the "POU instance", it may be necessary to specify the calling hierarchy, or even the program location in the case of functions, see Section 7.6.4.

As long as the PLC is kept in "forcing" mode by the programming system and with the parameters set as shown in Example 7.9, the program will always find the value Boolean 1 when reading %IX0.2, %IX0.6 or %IX0.14.

Depending on the implementation in the PLC, the variables being forced are either set to the forced value only once at the beginning of every cycle and the program itself can overwrite them during the cycle, or they are kept to the forced value throughout the cycle (overwriting is prevented).

7.6.6 Program test

Breakpoints and Single Step, functions well known from PC development environments, can also be used for debugging PLC programs:

- **Breakpoint Set/ Reset.** The user specifies a location in the program where execution is to be interrupted and further instructions from the user awaited. Advanced systems support conditional breakpoints, e.g. "Break at line 20 of block FB4 if Function_1 has been called, Variable_X has been set and Function_2 has been reached."

    As mentioned before, specifying the line number and POU or instance name is not always sufficient, but the location in the calling hierarchy should be specified, see Figure 7.5.

- **Single Step.** After stopping at a breakpoint, the user can execute the following instructions or graphical elements one at a time.

7.6.7 Testing Sequential Function Chart programs

Special features are required to test programs written in SFC. Depending on the system, the following features are available:

- Setting/Resetting transitions, to force or prevent a transition to the next step.
- Activation of steps or transitions, to begin execution of a POU at a specific location.
- Permanent blocking of steps or transitions, to prevent them from being activated.
- Reading and modifying system data that is relevant to SFC, e.g. step flags, step timers, etc.

This is the equivalent of forcing at SFC level.

7.7 Data Blocks and Recipes

A *recipe*, in PLC parlance, is a set of variable values, which can be replaced by a different one, to make the same PLC program behave differently. For example, in an automated process, only parameters like length, weight, or temperature need to be changed to produce a different product. Using recipes, such process data can be modified during operation.

The replacement of one set of data by another can be performed by the program itself. In this case, all possible data sets have to be stored in PLC memory. Another possibility is to download a new data set to the PLC at a convenient point during operation, replacing an existing data set.

Some PLC systems use *Data Blocks* (DB) for this purpose. A data block is accessible throughout the project (global) and consists of data items of specific data types.

The data block number gives the program the base address of the data block that is currently active. The data items are accessed by an index (data word number). To switch to a different data block, only the base address of the new block has to be selected in order to activate it.

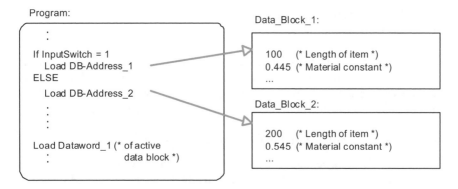

Figure 7.6. Using data blocks. Exactly one data block can be active at any one time.

In Figure 7.6, the instruction "Load Dataword_1" will load 100 or 200, depending on which data block is active.

Older (not IEC 61131-3-conformant) systems used data blocks also for (temporarily) storing their own local data.

Common features of data blocks can be summarised as follows:

- DBs can be downloaded and replaced separately like any other block,
- DBs are globally accessible by any other block,
- Exactly one DB is active at any one time,
- DBs contain data type information about their data items.

As the subject of data blocks has not been raised until this late stage of our book, it is only natural at this point to ask: "Where are the DBs in IEC 61131-3?"

IEC 61131-3 does not discuss data blocks. Instead of activating one global data block to hold parameters and local information for a function block, instantiation of function blocks is used (see Chapter 2). Each instance of a FB is automatically assigned its own "local" data record. FBs can also access global data. IEC 61131-3 therefore fully covers the functionality of data blocks as a means of assigning parameters to function blocks and storing local data.

However, unlike POUs, the instance data areas of FBs **cannot** be replaced or initialised separately. Sets of data such as recipes are either already included in the PLC program or replaced when the entire POU is replaced by a new one that contains the new data.

Ways of implementing most of the features of DBs using the methods of IEC 61131-3 are outlined here.

Data that belongs together is declared as a structure or an array. Like data blocks, these consist of a number of values, each having a data type. Such compound variables can be used as input or output parameters, local or global data. Data structures like this can also conveniently be used to implement recipes.

Switching between different sets of data can be done in several ways:

1) The PLC is stopped and a complete new program with different data is downloaded to the PLC. This method requires the PC to be permanently connected to the PLC and the process under control must be able to tolerate such interruptions.
2) Individual blocks (POUs) are replaced. The POUs downloaded from the PC replace POUs with identical names in the PLC, but have different initial values for the variables contained or call different POUs. The PLC's operating system must have a block management facility that allows replacement of individual blocks during operation.
3) Remote SCADA software is used to dynamically modify the (global) set of data.
4) All the sets of data required are contained within the PLC. The program invokes POUs with different sets of data or assigns its function blocks different parameter sets depending on the situation.

Replacing POUs seems appropriate for POUs which mainly:

- provide global data,
- copy or initialise data.

Global and external variables are a common way of supplying parameters to parts of a program, e.g. function blocks:

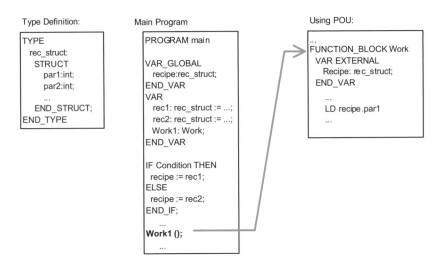

Type Definition:

```
TYPE
  rec_struct:
    STRUCT
      par1:int;
      par2:int;
      ...
    END_STRUCT;
END_TYPE
```

Main Program

```
PROGRAM main

VAR_GLOBAL
  recipe:rec_struct;
END_VAR
VAR
  rec1: rec_struct := ...;
  rec2: rec_struct := ...;
  Work1: Work;
END_VAR

IF Condition THEN
  recipe := rec1;
ELSE
  recipe := rec2;
END_IF;
  ...
Work1 ();
  ...
```

Using POU:

```
...
FUNCTION_BLOCK Work
  VAR EXTERNAL
    Recipe: rec_struct;
  END_VAR

  ...
  LD recipe.par1
  ...
```

Example 7.10. The main program contains different sets of data (rec1, rec2). The global structure recipe is assigned one of these depending on a specified condition. Other POUs simply access the global variable with a corresponding EXTERNAL declaration.

The disadvantage of using GLOBAL declarations is that all data sets have to be stored in the PLC all the time. Systems that support the configuration of global variables can avoid this problem by allowing a resource to be supplied with values for its global data by another resource or via access paths (VAR_ACCESS).

Global data is inherently error-prone when programs have to be modified, as side effects could occur in all locations where the data is accessed. The IEC 1131-3 principle of object-oriented data is also violated by using global variables. To prevent this, call parameters could be used instead of the global variables in Example 7.10.

Main Program

Using POU:

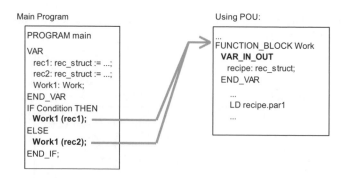

Example 7.11. To avoid the side effects that are possible when using the method in Example 7.10, data sets rec1 and rec2 are passed as input/ output parameters (this only passes a pointer and avoids copying huge data structures).

7.8 FB Interconnection

7.8.1 Data exchange and co-ordination of blocks in distributed systems

IEC 61131-3 defines PROGRAM blocks, which hierarchically call function blocks and functions, passing them parameters. Each PROGRAM (or certain function block instances) is assigned tasks. Communication between the tasks takes place using global variables or ACCESS variables.

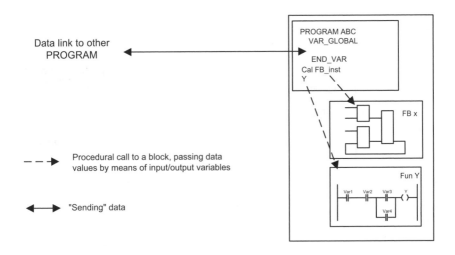

Figure 7.7. One program calling other blocks and passing them information through parameters. With other programs, only data is exchanged.

Distributed PLC systems, as used in industrial plants, power supply systems or building automation, require:

- parallel, autonomous execution of individual algorithms,
- geographically separate computing nodes,
- asynchronous data exchange.

Today, *distributed applications* are typically implemented as follows. Pre-fabricated blocks are copied from a library to create a new project. Missing functionality is implemented in new, specially written blocks. One PROGRAM, together with the function blocks and functions called, constitutes an executable unit.

Each PROGRAM is now assigned a node in the network (tasks of control units in the network), and the inputs and outputs of all programs are interconnected. Unconnected inputs of program instances are assigned individual parameter values where necessary. This is shown in Figure 7.8

Libraries are usually implemented by experts from the hardware manufacturer or are part of the firmware (EPROM) of the PLC or network node.

This is an extreme case of the IEC 61131-3 programming model. The application is "configured" from pre-fabricated blocks. The programs run mostly autonomously without being "called" by other blocks. Functions and function blocks in the sense of IEC 61131-3 are provided locally to be called by the PROGRAM. A direct call to a block in another node or CPU is not possible (a PROGRAM may not invoke a PROGRAM).

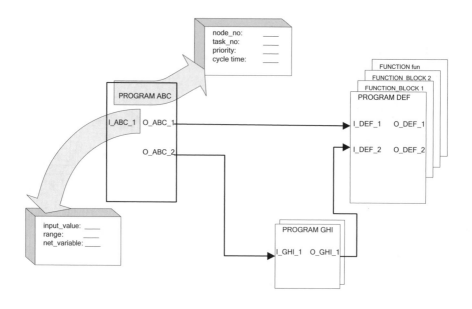

Figure 7.8. Assignment of blocks to network nodes and PLC tasks. All PROGRAMs run independently and are connected only via ACCESS variables.

It is possible to deviate from the definition in IEC 61131-3, which stipulates that only programs may be connected to objects in different tasks. Function blocks can then be distributed amongst computing nodes (tasks) and interconnected at will. This allows a much closer mapping of algorithms to network nodes. The performance of a distributed automation system can be increased without changing the program by simply adding computing nodes and re-configuring appropriately.

This involves two problems, as shown in Figure 7.8:

1) The run-time behaviour of all blocks must be properly co-ordinated because control information sent together with other data information is confusing.
2) A mechanism is needed to ensure consistent flow of data between blocks (network nodes) by checking the validity of data items or groups of data items.

As long as all blocks interconnected are executed on the same network node, execution control can be implemented implicitly (blocks are activated from top to bottom or from left to right, according to the function block network) or explicitly configured (execution number for every block). Additional control is needed if the tasks involved have no common time base, e.g. if they are executed on different nodes in the network.

These *interconnection techniques* result in even greater separation between programming and configuration. The programmer writes programs using FBs and functions and interconnects these. Afterwards, the function blocks are assigned to computing nodes. One and the same application can execute locally in one task, or be distributed between many tasks, without having to modify the program structure.

In essence, this method was picked up on in IEC 61499; the basic procedure is described in Chapter 9.

Communication blocks as described in IEC 61131-5 [IEC 61131-5] are another option for exchanging information between blocks of different configurations. They are provided by the manufacturer and can also be used for communicating with non-IEC 61131-3 systems.

7.8.2 Macro techniques in FB interconnection

A project with interconnections as described in Section 7.8.1 consists of a large number of blocks. To make the structure easier to understand, groups of blocks can be visually combined and shown as one block. Related functionality can be grouped onto separate working sheets, called *Function Charts*. This "*macro*" technique is explained below. Example 7.12 gives an example from the plant industry.

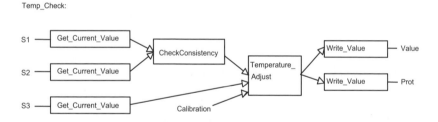

Example 7.12. Interconnection of simple basic elements in plant engineering

Placing and connection of blocks (Get_Current_Value, CheckConsistency, ...) is performed graphically by the user. All blocks have a clearly defined functionality. Data declarations are made implicitly by block instantiation. Only when connecting inputs and outputs does the user (or programming system) have to check for compatibility of data types.

Blocks provided by the manufacturer can be used to build more sophisticated blocks using a kind of macro technique. This corresponds to building complex data structures out of elementary data types. For this reason these blocks are sometimes called *Derived Function Blocks*.

Temp_Control:

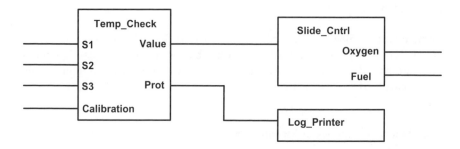

Example 7.13. The blocks from Example 7.12 (called Temp_Check) attain a higher degree of specialisation and sophistication when grouped together using macro techniques.

After definition as shown in Example 7.13, block Temp_Control can be used in applications.

If the blocks used are elementary blocks defined by the manufacturer (well-known behaviour), good simulation results can be achieved (e.g. run-time, effects of different hardware assignments or modifications in communication infrastructure).

7.9 Diagnostics, Error Detection and Error Handling

Diagnostics is basically the "detection of error conditions during operation and localisation of the source of error". There are four areas where errors can occur:

1) PLC hardware, including connections to other devices,
2) PLC software (operating system),
3) User software,
4) Process behaviour: the process under control may enter an unforeseen state.

A general distinction is made between system errors and programming errors.

Vendors offer various tools for diagnostics. These can be either additional hardware that checks for error conditions and provides information about them, or software functions to be included in the application. SFC is a good language for detecting errors in a running installation (e.g. "Transition XY not firing") or for continuing with a defined response after detecting an error.

Error concept of IEC 61131-3.

IEC 61131-3 has only a very general approach to error handling, giving the user a certain amount of support in handling cases 2) and 3) above. The standard requires an error list to be provided by PLC manufacturers, indicating the system response to a variety of specified error conditions (see Appendix E):

1) The error is not reported. There must be a statement to this effect in the user documentation.
2) The possibility that the error might occur is detected when preparing (editing, compiling or loading) the program and the user is warned accordingly.
3) The error is reported during execution of the program (at run time). Manufacturer-dependent procedures for handling the error are provided.

Quality assurance plays an important role in the automation business. The quality of today's compilers effectively prevents some typical program errors from slipping through at the compilation stage. Concepts of IEC 61131-3, like strict data type checking, even prevent some errors from occurring in the first place, during programming. However, some errors can only be detected at run time.

Some error situations, like division by zero (see Appendix E) should be checked by the PLC system. IEC 61131-4 (with supporting information for programmers [IEC 61131-4]) recommends the definition of a uniform global (manufacturer-dependent) data structure for errors, which should contain the status of an operation (Error Yes/ No), the type of error (Division by zero) and the location of the error (POU name). This information could then be scanned by the application, or connected to the SINGLE input of a task (see Section 6.3.4). This task would be connected to a system routine or application routine for specific error handling to initiate online correction, an error message to the user, a warm restart or similar.
    In the event of an error, the PLC system would set the error status to TRUE and set other members of the data structure accordingly, thus starting the error task.

Extended error handling model (beyond IEC).

To improve software quality, it is desirable to provide the users themselves with a means of defining error conditions in a standardised form. A language construct like "asserted conditions" could be used for this. In this case, the programmer would implement the application with checks.

For example:

- Is the value of a variable within the limits that apply at this program location?
- Do two variables match?

FBD:

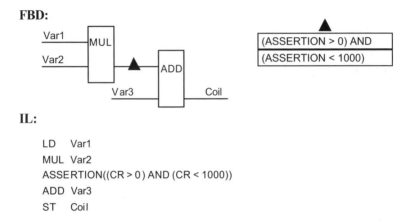

IL:

```
LD   Var1
MUL  Var2
ASSERTION((CR > 0 ) AND (CR < 1000))
ADD  Var3
ST   Coil
```

Example 7.14. For detection of run-time errors, it should be possible to implement checks that are calculated by the system itself. In the graphical languages, connections could be secured with assertion symbols. Expressions should be written in one of the languages of IEC 61131-3 (e.g. ST).

Such assertions can be simple expressions for the Current Result (CR) in IL, as shown in Example 7.14. Assertions can be used to check "critical" values at run time. Complex expressions can also be employed - at specified program locations - to compare input and output values, check for consistency or check important process parameters for logical relations.

Some systems provide automatic error checking facilities, e.g. for array indices, i.e. the index in an array must not be above the upper or below the lower limit of the array. This is supported by IEC 61131-3, see Chapter 3.

The response in the event of a violation of an assertion must be configurable, e.g.: Stop program; issue error message to visualisation system. For more sophisticated "exception handling", multiple error routines (user-defined or standard error POUs) should be assignable to different assertions. These routines should have special privileges, such as the right to stop or restart the PLC system.

At present, assertion conditions and the associated error responses (without special privileges) still have to be written by users themselves, which is not always an ideal solution from the point of view of program readability.

The architecture of error handling is evident throughout a program, and is at present dependent on the manufacturer. The lack of standardised error detection and error handling routines makes porting of applications between different systems difficult, requiring specially trained system experts.

7.10 Hardware Dependence

Studies have shown that even sophisticated cross-compilers can rarely automatically cross-compile more than 60% of a non-IEC 61131-3 PLC application to an IEC 61131-3 programming system. The reason is that the programs are heavily hardware-dependent. Custom routines are used to control special hardware, or specialised hardware addresses (status registers, system memory,...) are accessed.

IEC 61131-3 does not set out to eliminate the individuality of manufacturers. After all, a wide variety of software and hardware ensures high functionality. To ease portability, IEC 61131-3 provides the following mechanisms:

- All external information a program needs should be provided by IEC 61131-3-conformant global variables, access paths or communication FBs (see IEC 61131-5).
- Hardware I/O addresses used have to be declared in the PROGRAM or configuration scope.
- Hardware-dependent features have to be listed in a special table that manufacturers have to provide with their software.

The list of implementation-dependent parameters is given in Appendix F.

8 Main Advantages of IEC 61131-3

Chapter 1 outlines goals and benefits of IEC 61131-3 for manufacturers and users. How well does this programming standard live up to expectations?

Many features and concepts of this way of programming PLCs have been described and explained in previous chapters. The core concepts are summarised again here.

The following outstanding features of PLC programming with IEC 61131-3 deserve special notice:

- Convenience and security with variables and data types,
- Blocks with extended capabilities,
- PLC configuration with run-time behaviour,
- Uniform programming languages,
- Structured PLC programs,
- Trend towards open programming systems.

8.1 Convenience and Security with Variables and Data Types

Local and global variables instead of hardware addresses
Formerly, all data memory of a PLC was accessed using global addresses, and the programmer had to take care that one part of a program did not overwrite the data of another part. This applied particularly to I/O addresses, flags and data blocks.

IEC 61131-3 replaces all global hardware addresses by named variables with a defined scope: the programming system **automatically** distinguishes between global variables and variables local to a POU. Global addresses can be accessed by assigning the address to a named variable in the declaration part and using this variable in the program.

K.-H. John, M. Tiegelkamp, *IEC 61131-3: Programming Industrial Automation Systems*, 2nd ed., DOI 10.1007/978-3-642-12015-2_8,
© Springer-Verlag Berlin Heidelberg 2010

Type-oriented access to PLC data

PLC programmers used to have to be careful to use the same data type when reading or writing to individual PLC addresses. It was possible to interpret the same
memory location as an integer at one place in a program, and as a floating-point
number at another.

IEC 61131-3 prevents such programming errors from occurring, as each variable
(including direct hardware addresses) must be assigned a data type. The programming system can then check, that all accesses use the proper data type.

Defined initial values for user data

All data is explicitly declared in the form of a variable, and assigned a data type in
the declaration. Each data type has, either by default or as specified by the user, a
defined initial value, so each and every variable used in a program is always
correctly initialised in accordance with its properties.

Variables can be declared to be retentive (with a RETAIN qualifier). They are
then automatically assigned to a battery-backed area of memory by the
programming system.

Arrays and data structures for every application

Building on the predefined data types, the PLC programmer can design arrays and
other complex data structures to match the application, as is the practice with high-
level languages.

Limits of array indices and ranges of variable values are checked by the programming system as well as by the PLC system at run time.

Unified declaration of variables

The extensive facilities for using variables are generally identical in all the
languages defined by IEC 61131.

8.2 Blocks with Extended Capabilities

Reuse of blocks

Blocks (POUs), such as functions and function blocks, can be designed to be independent of the target system used. This makes it possible to have libraries of re-
usable blocks, available for multiple platforms.

Parameters of a function block instances, input as well as output, and local data
of each function block instance, keep their values between calls. Each instance of a
function block has its own data area in memory, where it can perform its
calculations independently of external data. It is not necessary to call a data block
for the FB to work on.

Programs can also be used in several instances and be assigned to different tasks
of one CPU.

Efficient assignment of block parameters
The standard provides a variety of mechanisms for passing data to and from blocks:

- VAR_INPUT: Value of a variable
- VAR_IN_OUT: Pointer to a variable (reference)
- VAR_OUTPUT: Return value
- VAR_EXTERNAL: Global variable of another POU
- VAR_ACCESS: Access path within a configuration.

Until now, the only items in this list that have been provided by most PLC systems have been global variables and the capability for passing values to a called block (but not for returning a value).

Standardised PLC functionality
To standardise typical PLC functionality, IEC 61131-3 defines a set of standard functions and function blocks. The calling interface, the graphical layout and the run-time behaviour of these is strictly defined by the standard.

This standard "library" for PLC systems is an important foundation for uniform and manufacturer-independent training, programming and documentation.

8.3 PLC Configuration with Run-Time Behaviour

Configurations structure PLC projects
Tasks and programs are assigned to the controller hardware at the highest level of a PLC project (the configuration). This is where the run-time properties, interfaces to the outside, PLC addresses and I/Os are defined for the various program parts.

Run-time features for PLC programs
Until now, the methods of specifying run-time properties, like cycle time and priority of programs, have often been system-specific. With IEC 61131-3, such parameters can be specified and documented individually by defining tasks.

PLC programs must not be recursive. The amount of memory required to hold the program at run time can therefore be determined off-line, and the programs are protected from unintentional recursion.

8.4 Uniform Programming Languages

IEC 61131-3 defines five programming languages, which can cover a wide range of applications.

As a result of this international standard, PLC specialists will in future receive more uniform training, and will "speak the same language" wherever they are employed.

The cost of training will be reduced, as only specific features of each new controller system have to be learned.

Documentation will be more uniform, even if hardware from more than one vendor is being used.

8.5 Structured PLC Programs

The various language elements of IEC 61131-3 allow clear structuring of applications, from definitions of blocks and data up to the hardware configuration.

This supports structured programming (top-down and bottom-up) and facilitates service and maintenance of applications.

The "structuring language", Sequential Function Chart, also enables users to formulate complex automation tasks clearly and in an application-oriented way.

8.6 Trend towards Open PLC Programming Systems

Standardisation of programming languages leads to standardisation of software, which makes vendor-independent, portable programs feasible, as is, for example, already the case in the personal computer domain with programming languages like Assembler, PASCAL, COBOL and, most notably, C. The "feature tables" of IEC 61131-3 provide a basis for comparing programming systems with the same basic functionality from different vendors. There will still be differences between the systems of different manufacturers, but these will mainly be found in additional tools like logic analysers or off-line simulation, rather than in the programming languages themselves.

The common look and feel in PLC programming is becoming more and more international and bring the separate markets in Europe, the US or Asia closer together.

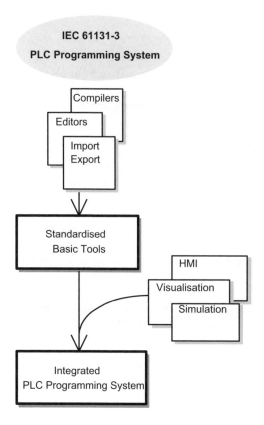

Figure 8.1. Trend towards open, standardised components built on IEC 61131-3-compliant programming systems

The new generation of PLC programming systems will have a standardised basic functionality, plus highly sophisticated additional tools to cover a wide range of applications.

As shown in Figure 8.1, standardisation by IEC 61131-3 promotes integrated systems, built from standardised components like editors, compilers, export and import utilities, with open, re-usable interfaces.

Tools which are traditionally sold as separate packages, like HMI, simulation or visualisation, will cause de facto interface standards to be established for these components.

8.7 Conclusion

IEC 61131-3 will cause PLC manufacturers and users to give up old habits for a new, state-of-the-art programming technology. A comprehensive standard like IEC 61131 was necessary to achieve a uniform environment for innovating the configuration and programming of PLC systems.

This manufacturer-independent standard will reduce the training and familiarisation time for PLC programmers, programs written will be more reliable, and the functionality of PLC systems will catch up with the powerful software development environments available for PCs today.

Compliance with IEC 61131-3, migration paths from legacy systems towards the new architectures, and a powerful, ergonomic user interface will be the most important criteria for users for a wide acceptance of the new generation of PLC programming systems.

Today's complex requirements and economic constraints will lead to flexible, open, and therefore manufacturer-independent PLC programming systems.

9 Programming by Configuring with IEC 61499

Programming using graphical elements taken from the "real world" of the application to be programmed is becoming more and more important.

With the graphical languages LD, FBD or SFC of IEC 61131-3 previously discussed, data flow and logical execution sequence can be programmed and documented using symbols and names. However, it is also desirable to be able to display the topological distribution of programs, their general configuration and interconnections to other parts of a distributed automation project in a graphical manner. This takes place at a higher, more abstract level than the programming of POUs described so far.

The tools for configuring complex and distributed applications are called *configuration editors*. Program parts, such as function blocks, are combined to form larger units. This is done by *interconnection* of function blocks.

In order to standardise unified language elements for this purpose, international standard *IEC 61499* was defined. It can be considered both a separate standard and a supplement to the existing IEC 61131. This chapter gives a summary of the basic concepts and ideas of this additional standard and explains its relationship with IEC 61131. The subject of standard IEC 61499 would need to be discussed in greater detail in another book, see also [IEC 61499-1], [IEC 61499-2] and [IEC 61499-4].

9.1 Programming by FB Interconnection with IEC 61131-3

In order to clarify the differences between IEC 61499 and IEC 61131-3, we shall first look at some special features of distributed programming.

The programming languages described in Chapter 4 are used to define algorithms for blocks. Function blocks and functions call each other, exchange information by means of their parameters and form a program in conjunction with a POU of type PROGRAM. A program runs as a task on a resource (CPU of a PLC).

K.-H. John, M. Tiegelkamp, *IEC 61131-3: Programming Industrial Automation Systems*, 2nd ed., DOI 10.1007/978-3-642-12015-2_9,
© Springer-Verlag Berlin Heidelberg 2010

IEC 61131-3 essentially concentrates on describing single programs together with their execution conditions. Information exchange between programs takes place using ACCESS variables or global data areas. This topic is discussed in Section 7.8 and illustrated by Figure 7.8.

Complex, distributed automation tasks have an extensive communication and execution structure. Intensive data exchange takes place between geographically separate control units. The semantic and temporal dependencies and conditions have to be specified.

To do this, programs are assigned to tasks of network nodes, execution conditions are defined as described in Section 6.1, and the inputs and outputs of programs (such as network addresses or parameter values) are interconnected.

Creating distributed automation solutions, i.e. configuring function blocks for physically different and geographically separate hardware and synchronising their execution, is the subject of standard IEC 61499. Although the concept emphasises the distribution to several automation devices, it may also be helpful to run an IEC 61499 application on a single control unit.

9.2 IEC 61499 – The Programming Standard for Distributed PLC Systems

The sequential invocation of blocks defined in IEC 61131-3 is not a suitable method for program structuring in distributed systems. This is already apparent in Figure 7.8. The goal of a distributed, decentralised system is to distribute programs between several control units and to execute them in parallel (in contrast to sequential execution with invocation by CAL). Here it is essential to ensure data consistency between nodes of the networked system, i.e. to define exact times for mutual data exchange.

Two kinds of information exchange play an essential part in IEC 61499:

1) *Data flow* of user data,
2) *Control flow*, which controls the validity of user data as event information.

The interaction of data and control flow can also be programmed by means of IEC 61131-3 using global variables and access paths. But the resulting overall program can easily become hard to read and slower to execute.

In order to describe the interactions between program parts and elements of control hardware within a distributed, networked automation system easily and exactly, IEC 61499 uses a model ("top-down" approach) with several hierarchical levels:

- System
- Device
- Resource

- Application
- Function block

The definitions of the terms *Resource* and *Function Block* are, however, wider than those of IEC 61131-3, as will be explained in this chapter.

Instead of assigning PROGRAM and TASK to a resource, function blocks in IEC 61499 can be assigned *run-time properties* **directly** via the resource .

9.2.1 System model

In a real automation environment several control units, referred to here as *devices*, execute the same or different applications in parallel. This is outlined in Figure 9.1.

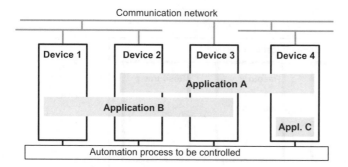

Figure 9.1. Control of a real process can be **distributed** between several devices. As in IEC 61131-3, several programs can also be configured for one device. The program parts interchange information via communication networks.

9.2.2 Device model

Closer examination of a device, as in Figure 9.2, shows that it consists of:

- its application programs,
- an interface to the communication network,
- an interface to the automation process,
- the device hardware, on which the resources run.

A *resource* represents an independent executable unit with parameters (a task in the general sense). Several resources can run on each device, and they can perform the same or different applications.

IEC 61499 uses two views of a distributed program, which are explained in this chapter. On the one hand, this standard looks at the hierarchy of System—Device—Resource, in order to describe system structure and the corresponding run-time properties. On the other hand, it also defines the user-oriented view of a distributed program. This user view is summarised by application and function block models that are discussed later in this chapter.

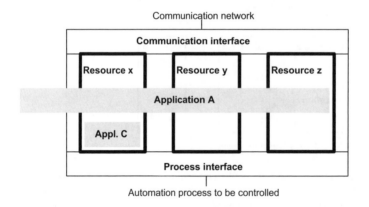

Figure 9.2. A device can contain several resources, which use common interfaces to exchange information with other control units and the automation process.

9.2.3 Resource model

A resource consists of function *blocks*, which exchange event and data information using special interfaces. There are two kinds of function blocks:

1) *Service interface function blocks*, which are standard FBs and form the interfaces to the automation process and the communication network.
2) User-defined function blocks, which make up the actual application program (algorithm).

As in IEC 61131-3, there is a distinction between FB type and FB instance.

Run-time properties, such as the maximum number of instances, execution time, number of connections etc., can be assigned to each function block within the resource.

The interconnection of FBs by the user is not carried out at resource level, but at application level (see next section). The real information exchange between the FBs of the application program takes place "invisibly" for the user via the communication and process interfaces.

An application can be implemented on one or more resources.

9.2.4 Application model

This section deals with the user-oriented view of a program. This view corresponds to the horizontal, grey "Application" bar in Figure 9.3, which can extend over several devices or resources.

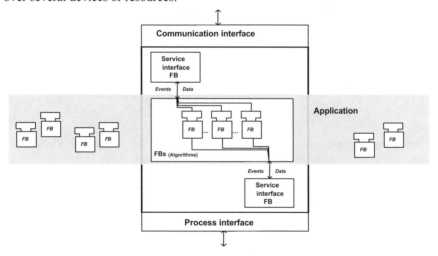

Figure 9.3. A resource consists of function blocks for controlling and data processing (algorithms) together with interface blocks (communication/process).

The application level forms the real programming level because it is here that the FBs are interconnected with one another, independently of the resources on which they run. It describes the application – which may subsequently be distributed amongst several resources.

After the application program, consisting of several FBs, has been assigned to the resources and the program has been started, communication takes place implicitly or explicitly (depending on the vendor) via the service interfaces with the connections specified by the user.

Figure 9.4 shows how an application is made up of both controlling parts (with events) and data processing parts (algorithms). Here the different graphical representation to that of IEC 61131-3 can be seen.

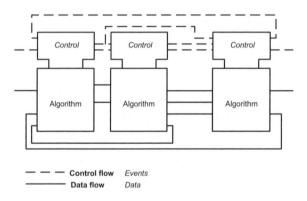

Figure 9.4. The application consists of interconnected function blocks; each of which has both controlling (control flow) and data processing (data flow) functions.

Control and data information always flows into a function block from the left and is passed on after processing from the outputs on the right.

9.2.5 Function block model

The function blocks are the smallest program units (like POUs). Unlike the FBs of IEC 61131-3, a function block in IEC 61499 generally consists of two parts:

1) Execution control: Creation and processing of events with control inputs and outputs (*control flow*),
2) Algorithm with data inputs and outputs and internal data (*data flow* and processing).

These function blocks can be specified in textual or graphical form. Function blocks are instantiated for programming, as in IEC 61131-3. The language elements for FB interface description are therefore very similar, see also Chapter 2.

Figure 9.5 shows the graphical representation of a function block in accordance with IEC 61499.

The algorithm part is programmed in IEC 61131-3 (like a POU body).
  The execution control part is programmed using a state diagram or sequential function chart (SFC in IEC 61131-3). The events are input values for *state diagrams*, or *execution control charts (ECC)* . These ECCs control the execution times of the algorithm or parts of it depending on the actual state and incoming events.

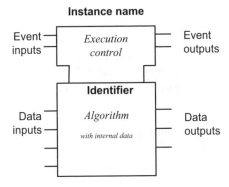

Figure 9.5. Graphical representation of a function block. Details of execution control, the internal algorithm and internal data are not shown at this level.

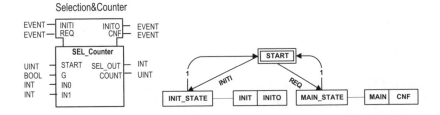

Example 9.1. Function block with typed formal parameters and state diagram (ECC)

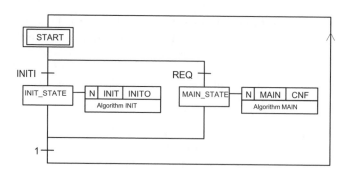

Example 9.2. Execution control of Example 9.1 using Sequential Function Chart (SFC) as defined in IEC 61131-3. The output events CNF and INITO are set by the application program and controlled by calling standard function blocks.

Example 9.1 contains function block SEL_Counter (instance name Selection&Counter), which consists of an ECC control part and the algorithm part, which in turn consists of the two algorithms INIT and MAIN. The execution control determines which algorithm part will be active at what time.

In Example 9.1, when the event INITI occurs, the FB control will change from initial state START to state INIT_STATE and algorithm INIT will be executed. Afterwards event output variable INITO is set (action "N"), followed by a RESET (i.e. a signal pulse). Now execution control evaluates the next transition. This has the constant parameter "1" in this example, which means the condition is always true, and leads back to state START. Incoming event REQ is processed analogously.

This behaviour is equivalent to the actions of Sequential Function Chart (SFC) in IEC 61131-3 and is illustrated by Example 9.2. IEC 61499 assumes that it is more favourable to specify the execution control using state diagrams, as with this method only **one** state can be active at a time. In SFC this can be achieved by prohibiting simultaneous branches.

Example 9.3 shows the textual definition of the FB type in Example 9.1.

The keyword WITH connects an event input/output with a data input/output. If an event parameter is set, it indicates the validity of the corresponding data line (assigned by WITH).

```
FUNCTION_BLOCK SEL_Counter
    EVENT_INPUT
            INITI  WITH START;              ALGORITHM INIT:
            REQ  WITH G, IN0, IN1;
    END_EVENT                                   INTERNAL_COUNT := START;

    EVENT_OUTPUT                                COUNT := INTERNAL_COUNT;
            INITO  WITH COUNT;
            CNF  WITH SEL_OUT, COUNT;       END_ALGORITHM
    END_EVENT

    VAR_INPUT                               ALGORITHM MAIN:
            START:          UINT;
            G:              BOOL;              IF G = 0 THEN SEL_OUT := IN0;
            IN0, IN1:       INT;               ELSE
    END_VAR                                        SEL_OUT := IN1;
                                               END_IF;
    VAR_OUTPUT
            SEL_OUT:        INT;                INTERNAL_COUNT :=
            COUNT:          UINT;                 INTERNAL_COUNT +1;
    END_VAR
                                               COUNT := INTERNAL_COUNT;
    VAR
            INTERNAL_COUNT: UINT;           END_ALGORITHM
    END_VAR

    ...
    END_FUNCTION_BLOCK
```

Example. 9.3. Example 9.1 in textual representation (Structured Text ST of IEC 61131-3).

Composite function blocks

For the purposes of clear, object-oriented representation, several *basic function blocks* can be combined to form a new *composite function block*, which looks just like a "normal" function block on the outside, as shown in Figure 9.6.

Composite function blocks do not have their own execution control part, as this is the sum of the controls of all basic FBs of which it is composed. In the graphical representation in Figure 9.6 a) the FB header is therefore "empty".

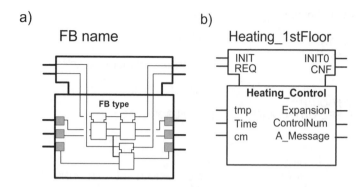

Figure 9.6. A composite function block consists of several interconnected function blocks with a common interface.
a) Example: Internal structure of a composite FB,
b) Example: External appearance of this FB.

9.2.6 Creating an application

IEC 61499 application programmers write programs by configuring and assigning parameters to ready-made function blocks.

FB	Explanation
Standard FBs	• FBs with functionality as in IEC 61131-3 • Service interface FBs (standardised communication services) • Event FBs (standardised event generation and processing)
User-defined FBs	Algorithms and ECC execution control e.g. programmed with IEC 61131-3

Table 9.1. Different types of function blocks in IEC 61499

The function blocks in Table 9.1 can be implemented as basic or composite FBs with a uniform interface.

For example, Event FBs provide functions for merging and splitting events or creating single or cyclic events.

A *configuration editor* is used for allocating blocks to resources (devices) and for interconnecting FBs.

9.3 Overview of the Parts of IEC 61499

The standard IEC 61499 consists of three parts, whose main contents are summarised in Table 9.2 ([IEC 61499-1 to -4]) (as of May 2008). A revised version of this standard is envisaged for 2010.

Parts	Contents
1. *Architecture*	Introduction and modelling, describes the validity, defines common terms, specification of function blocks, service interfaces, configuration and syntax.
2. *Software Tool Requirements*	Contains descriptions to support the life cycle of distributed programs. This document is still in the design phase.
3. *Rules for Compliance Profiles*	Describes rules for checking the compatibility of user programs with the standard, similarly to IEC 61131-3.

Table 9.2. Structure and contents of standard IEC 61499

10 Contents of CD-ROM and DVD

10.1 IEC Programming Systems STEP 7 and OpenPCS

The DVD and CD-ROM enclosed in this book contain the following information, examples and programs:

1) **DVD: STEP 7 Professional 2006 SR5 Engineering Software for SIMATIC S7 / M7 / C7** [1] as a system for PLC programming with IEC 61131-3 using the languages: IL, KOP, FUP, S7-GRAPH and S7-SCL as well as PLC simulation PLCSIM and the add-on iMAP. The package is completed with the Automation License Manager and electronic manuals; running under Windows XP Professional SP3, Windows Server 2003 SP2 standard edition and Windows Vista 32 Bit Ultimate/Business with/without SP1.

2) **CD-ROM: Open PCS**, a system (full version) for programming with IEC 61131-3 , running on any standard Windows PC, using the languages: IL, LD, FBD, SFC, ST and CFC[2]; running under Windows Server 2003, Windows XP SP2 or Windows Vista 32bit. PLC simulation SmartPLC is available for simulating the programs on a PC. The dedicated OPC server SmartPLC/OPC is only required, if additional third-party hardware and/or external OPC clients are connected.

3) **CD-ROM: Examples** of this book as source files,

4) **CD-ROM: Buyer's Guide** for IEC 61131-3-compliant programming systems.

File README.TXT on the CD contains important information about the installation and use of the files and programs. It shows how to copy the files onto hard disk and gives tips on how to use the examples and the buyer's guide.

README.TXT is an ASCII file and can be read using any text editor.

The files of the two programming systems are either self-extracting or in the form of an installation package, i.e. they cannot be read immediately but must first be decompressed or installed. No additional software is needed.

---

[1] IL corresponds to STL, FUP to FBS, KOP to LD, S7-GRAPH to SFC, and S7-SCL to ST

[2] CFC: not part of IEC 61131-3

K.-H. John, M. Tiegelkamp, *IEC 61131-3: Programming Industrial Automation Systems*, 2nd ed., DOI 10.1007/978-3-642-12015-2_10,
© Springer-Verlag Berlin Heidelberg 2010

Demo versions of STEP 7 (Siemens, www.siemens.com) and OpenPCS (infoteam, www.infoteam.de)

With the aid of the above-mentioned versions of two selected programming systems, readers can program, modify, extend and test all examples in this book or create programs of their own in order to practise PLC programming with IEC 61131-3.

STEP 7 Professional includes the programming languages as well as a PLC simulation S7-PLCSIM, which allows testing of programs without physical hardware. OpenPCS contains all IEC 61131-3 programming languages, the programming language CFC as well as a run-time package Smart PLC/SIM, which additionally allows execution of a PLC program on PC (offline simulation), and the OPC server SmartPLC/OPC.

Hints on using and purchasing commercially available software versions of both programming systems (as well as hardware) can also be found in the relevant folders on the CD and/or DVD.

The authors are not responsible for the contents and correct functioning of these programming versions. These software packages have a partly restricted scope compared with the functionality of the corresponding products. Their use is only allowed in conjunction with this book for learning and training purposes.

IL examples

To save the reader having to re-type the programming examples in this book, the most important IL examples are provided on the CD. Further information about these can also be found in README.TXT.

10.2 Buyer's Guide for IEC 61131-3 PLC Programming Systems

The CD-ROM also contains a buyer's guide as a file in the format "Word for Windows (1997-2003, 2007)".
Contents of the buyer's guide (file **BuyGuide.doc**):

Buyer's Guide for IEC 61131-3 PLC Programming Systems

    Checklists for evaluation of PLC programming systems
    Using the checklists
    Checklists for PLC programming systems

        Compliance with IEC 61131-3
        Language scope, decompilation and cross-compilation
        Tools
        Working environment, openness, documentation
        General, costs

This buyer's guide essentially consists of tables, or "checklists", which permit objective evaluation of PLC programming systems compliant with the standard IEC 61131-3. The use of these lists is described in detail before explaining each criterion for PLC programming systems.

The file can be copied for multiple product evaluation and individual editing of the checklists. These tables are stored on the CD in three different file formats:

1) Microsoft Word for Windows (1997-2003, 2007) (file TABLES.DOC),
2) Microsoft Excel for Windows (1997-2003, 2007) (file TABLES.XLS),
3) RTF text (file TABLES.RTF).

The Excel version is advantageous, as all calculations can be done automatically.

A Standard Functions

This appendix contains a complete overview of all the standard PLC functions described by means of examples in Chapter 2. For every standard function of IEC 61131-3, the following information is given:

- Graphical declaration
- (Semantic) description
- Specification of some functions in Structured Text (ST).

Standard functions have input variables (formal parameters) as well as a function value (the returned value of the function). Some input variables are not named. In order to describe their functional behaviour the following conventions apply:

- An individual input variable without name is designated as "IN"
- Several input variables without names are numbered "IN1, IN2, ..., INn"
- The function value is designated as "F".

General data types (such as ANY or ANY_BIT) are used in the description. Their meaning is explained in Section 3.4.3 and they are summarised in Table 3.9. The designator ANY here stands for one of the data types: ANY_BIT, ANY_NUM, ANY_STRING, ANY_DATE or TIME.

Many standard functions have a textual name as well as an alternative representation with a symbol (e.g. ADD and "+"). In the figures and tables, both versions are given.

K.-H. John, M. Tiegelkamp, *IEC 61131-3: Programming Industrial Automation Systems*, 2nd ed., DOI 10.1007/978-3-642-12015-2,
© Springer-Verlag Berlin Heidelberg 2010

A.1 Type Conversion Functions

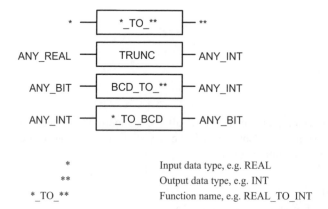

*	Input data type, e.g. REAL	
**	Output data type, e.g. INT	
*_TO_**	Function name, e.g. REAL_TO_INT	

Figure A.1. Graphical declarations of the type conversion functions

These standard functions convert the input variable into the data type of their function value (type conversion).

Name	Function / Description
*_TO_**	When REAL values are converted to INT values, they are rounded up or down to the next whole number. Halves, e.g. 0.5 or 0.05, are rounded up.
TRUNC	This function cuts off the places of a REAL value after the decimal point to form an integer value.
BCD	The input and/or output values of type ANY_BIT represent BCD-coded bit strings for the data types BYTE, WORD, DWORD and LWORD. BCD coding is not defined by IEC 61131-3, it is implementation-dependent.

Table A.1. Description of the type conversion functions

A.2 Numerical Functions

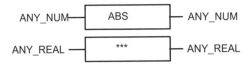

ANY_NUM—	ABS	— ANY_NUM
ANY_REAL —	***	— ANY_REAL

*** stands for: SQRT, LN, LOG, EXP,
                    SIN, COS, TAN,
                    ASIN, ACOS, ATAN

Figure A.2. Graphical declarations of the numerical functions

Name	Function	Description
ABS	Absolute value	$F := |IN|$
SQRT	Square root	$F := \sqrt{IN}$
LN	Natural logarithm	$F := \log_e(IN)$
LOG	Logarithm base 10	$F := \log_{10}(IN)$
EXP	Exponent base e	$F := e^{IN}$
SIN	Sine, IN in radians	$F := SIN(IN)$
COS	Cosine, IN in radians	$F := COS(IN)$
TAN	Tangent, IN in radians	$F := TAN(IN)$
ASIN	Principal arc sine	$F := ARCSIN(IN)$
ACOS	Principal arc cosine	$F := ARCCOS(IN)$
ATAN	Principal arc tangent	$F := ARCTAN(IN)$

Table A.2. Description of the numerical functions

A.3 Arithmetic Functions

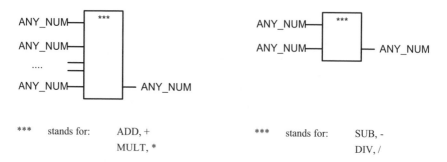

Figure A.3. Graphical declarations of the arithmetic functions ADD, MUL, SUB and DIV

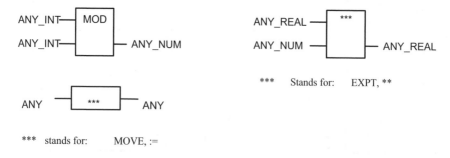

Figure A.4. Graphical declarations of the standard arithmetic functions MOD, EXPT and MOVE

Name	Symbol	Function	Description
ADD	+	Addition	$F := IN1 + IN2 + ... + INn$
MUL	*	Multiplication	$F := IN1 * IN2 * ... * INn$
SUB	-	Subtraction	$F := IN1 - IN2$
DIV	/	Division	$F := IN1 / IN2$
MOD		Remainder formation	$F := IN1 - (IN1/ IN2)*IN2$
EXPT	**	Exponentiation	$F := IN1^{IN2}$
MOVE	:=	Assignment	$F := IN$

Table A.3. Description of the arithmetic functions

ADD and SUB use the type ANYMAGNITUDE, which includes operations on TIME. EXPT uses the type ANY_REAL for IN1 and the type ANY_NUM for IN2. MOVE uses ANY for input and output.

In the case of the division of integers, the result must also be an integer. If necessary, the result is truncated in the direction of zero.

If the input parameter IN2 is zero, an error is reported at run time with error cause "division by zero", see also Appendix E.

A.4 Bit-Shift Functions

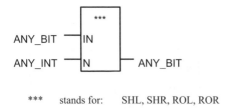

***   stands for: SHL, SHR, ROL, ROR

Figure A.5. Graphical declarations of the bit-shift functions SHL, SHR, ROR and ROL

Name	Function	Description
SHL	Shift to the left	Shift IN to the left by N bits, fill with zeros from the right
SHR	Shift to the right	Shift IN to the right by N bits, fill with zeros from the left
ROR	Rotate to the right	Shift IN to the right by N bits, in a circle
ROL	Rotate to the left	Shift IN to the left by N bits, in a circle

Table A.4. Description of the bit-shift functions. Input N must be greater than or equal to zero.

A.5 Bitwise Boolean Functions

*** stands for: AND, &, OR, >=1, XOR, =2k+1

Figure A.6. Graphical declarations of the bitwise Boolean functions AND, OR, XOR and NOT

Name	Symbol	Function	Description
AND	&	Bit-by-bit AND	F := IN1 & IN2 & ... & INn
OR	>=1	Bit-by-bit OR	F := IN1 v IN2 v ... v INn
XOR	=2k+1	Bit-by-bit XOR	F := IN1 XOR IN2 XOR ... XOR INn
NOT		Negation	F := ¬ IN

Table A.5. Description of the bitwise Boolean functions

The logic operation is performed on the input parameters bit-by-bit. That is, every bit position of one input is gated with the corresponding bit position of the other input and the result is stored in the same bit position of the function value.

An inversion can also be represented graphically by a circle "o" at the Boolean input or output of a function.

A.6 Selection Functions for Max., Min. and Limit

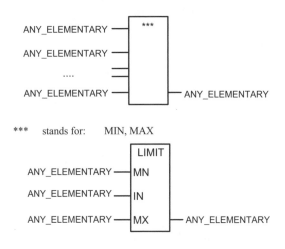

*** stands for: MIN, MAX

Figure A.7. Graphical declarations of the selection functions MAX, MIN and LIMIT

Name	Function	Description
MAX	Maximum formation	F := MAX (IN1, IN2, ... , INn)
MIN	Minimum formation	F := MIN (IN1, IN2, ... , INn)
LIMIT	Limit	F := MIN (MAX (IN, MN), MX)

Table A.6. Description of the selection functions MAX, MIN and LIMIT

These three standard functions are specified by declarations in ST in Example A.1 and Example A.2.

```
FUNCTION        MAX : ANY_ELEMENTARY  (* maximum formation *)
  VAR_INPUT     IN1, IN2, ... INn   : ANY_ELEMENTARY;        END_VAR
  VAR           Elem              : ANY_ELEMENTARY;          END_VAR
                (* ANY_ELEMENTARY  stands for any data type*)


  IF IN1 > IN2 THEN        (* first comparison *)
     Elem    := IN1;
  ELSE
     Elem    := IN2;
  END_IF;
  IF IN3 > Elem THEN       (* next comparison *)
     Elem    := IN3;
  END_IF;
  ...
  IF INn > Elem THEN       (* last comparison *)
     Elem    := INn;
  END_IF;
  MAX     :=  Elem;        (* writing the function value *)
END_FUNCTION
```

Example A.1. Specification of the selection function MAX in ST; for MIN replace all „>"
with „<".

The specification of the MIN function can be obtained by replacing all occurrences
of ">" by "<" in the MAX specification in Example A.1.

```
FUNCTION        LIMIT : ANY_ELEMENTARY (* limit formation *)
  VAR_INPUT
    MN   :   ANY_ELEMENTARY;
    IN   :   ANY_ELEMENTARY;
    MX   :   ANY_ELEMENTARY;
  END_VAR
  MAX    :=  MIN ( MAX ( IN, MN), MX);        (* call of MIN of MAX *)
END_FUNCTION
```

Example A.2. Specification of the selection function LIMIT in ST

A.7 Selection Functions for Binary Selection and Multiplexers

Figure A.8. Graphical declarations of the selection functions SEL and MUX

Name	Function	Description
SEL	Binary selection	F := IN0, if G = 0, otherwise IN1
MUX	Multiplexer	F := INi, if K = i and 0<=K<=n-1

Table A.7. Description of the selection functions SEL and MUX

These two standard functions are specified in the following examples by declaration in ST:

```
FUNCTION      SEL : ANY         (* binary selection *)
   VAR_INPUT
     G    :   BOOL;
     IN0  :   ANY;
     IN1  :   ANY;
   END_VAR
   IF G = 0 THEN
       SEL   :=  IN0;      (* selection of upper input *)
   ELSE
       SEL   :=  IN1;      (*selection of lower input *)
   END_IF;
END_FUNCTION
```

Example A.3. Specification of the selection function SEL in ST

```
FUNCTION       MUX : ANY          (* multiplexer *)
  VAR_INPUT
    K    :   ANY_INT;
    IN0  :   ANY;
    IN1  :   ANY;
    ...
    Inn-1 :   ANY;
  END_VAR
  IF (K < 0) OR (K>n-1) THEN
      ... error message .... ;          (* K negative or too large *)
  END_IF;
  CASE  K  OF
      0: MUX := IN0;          (* selection of upper input *)
      1: MUX := IN1;          (* selection of second input *)
      ...
      n-1: MUX := INn;        (* selection of lowest input *)
  END_CASE;
END_FUNCTION
```

Example A.4. Specification of the selection function MUX in ST

A.8 Comparison Functions

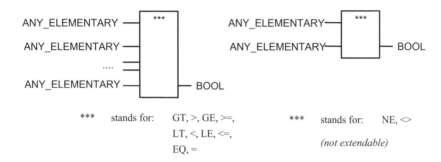

*** stands for: GT, >, GE, >=,
LT, <, LE, <=,
EQ, =

*** stands for: NE, <>

(not extendable)

Figure A.9. Graphical declarations of the comparison functions GT, GE, LT, LE, EQ, NE

Name	Function	Description
GT	Comparison for „> "	F := 1, if INi > IN(i+1), otherwise 0
GE	Comparison for „> ="	F := 1, if INi >= IN(i+1), otherwise 0
LT	Comparison for „< "	F := 1, if INi < IN(i+1), otherwise 0
LE	Comparison for „< ="	F := 1, if INi <= IN(i+1), otherwise 0
EQ	Comparison for „= "	F := 1, if INi = IN(i+1), otherwise 0
NE	Comparison for „<>"	F := 1, if INi <> IN(i+1), otherwise 0

Table A.8. Description of the comparison functions

The specification of these standard functions is illustrated in Example A.5 by the declaration of GT in ST, from which the others can easily be derived.

```
FUNCTION      GT : BOOL        (* comparison for 'greater than' *)
  VAR_INPUT    IN1, IN2, ... INn  : ANY_ELEMENTARY;       END_VAR
  IF      (IN1 > IN2)
  AND    (IN2 > IN3)
  ...
  AND    (IN(n-1) > INn) THEN
     GT :=  TRUE;         (* inputs are "sorted": in increasing monotonic sequence *)
  ELSE
     GT :=  FALSE;        (* condition not fulfilled *)
  END_IF;
END_FUNCTION
```

Example A.5. Specification of the comparison function GT in ST

A.9 Character String Functions

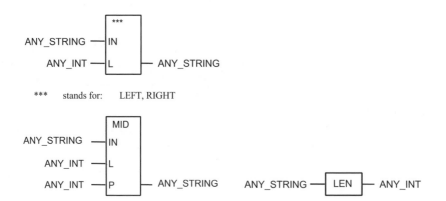

***   stands for: LEFT, RIGHT

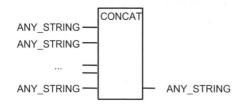

Figure A.10. Graphical declarations of the character string functions LEFT, RIGHT, MID and LEN

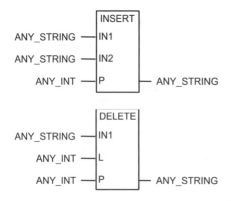

Figure A.11. Graphical declaration of the character string function CONCAT

Figure A.12. Graphical declarations of the character string functions INSERT and DELETE

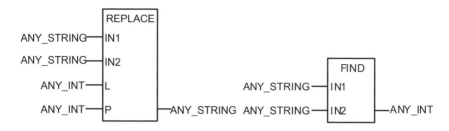

Figure A.13. Graphical declarations of the character string functions REPLACE and FIND

Name	Function	Description
LEN	Determines the length of a character string	F := number of the characters in IN
LEFT	Starting section of a character string	F := starting section with L characters
RIGHT	Final section of a character string	F := final section with L characters
MID	Central section of a character string	F := middle section from position P with L characters
CONCAT	Sequence of character strings	F := total character string
INSERT	Inserts one character string into another one	F := total character string with new part from position P
DELETE	Deletes section in a character string	F := remaining character string with deleted part (L characters) from position P
REPLACE	Replaces a section of a character string with another one	F := total character string with replaced part (L characters) from position P
FIND	Determines the position of a section in a character string	F := index of the found position, otherwise 0

Table A.9. Description of the character string functions

The positions of the characters within a character string of type ANY_STRING are numbered starting with position "1".

Inputs L and P must be greater than or equal to zero.

A.10 Functions for Time Data Types

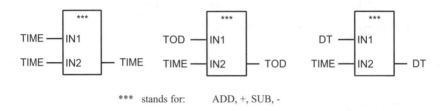

*** stands for: ADD, +, SUB, -

Figure A.14. Graphical declarations of the common functions for addition and subtraction of time

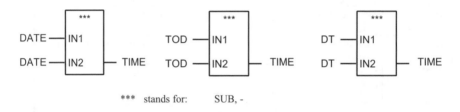

*** stands for: SUB, -

Figure A.15. Graphical declarations of the additional functions for subtraction of time

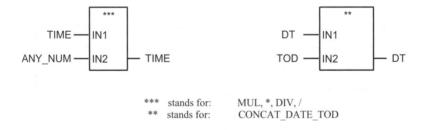

*** stands for: MUL, *, DIV, /
** stands for: CONCAT_DATE_TOD

Figure A.16. Graphical declarations of the functions MUL, DIV and CONCAT_DATE_TOD for time

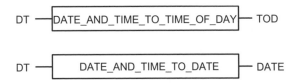

Figure A.17. Graphical declarations of the type conversion functions for time

The abbreviations TOD and DT can be employed as equivalents for the longer keywords TIME_OF_DAY and/or DATE_AND_TIME.

The function name versions ADD_TIME, SUB_TIME and the combinations ADD_TOD_TIME, ADD_DT_TIME, SUB_DATE_DATE, SUB_TOD_TIME, SUB_TOD_TOD, SUB_DT_TIME, MULTIME, DIVTIME and CONCAT_DATE_TOD can be used as equivalents for the function names ADD and SUB.

The standard explicitly advises against using the arithmetic symbols "+", "-", "/" and "*" and the function names ADD, SUB, DIV and MUL to prevent confusion and, as a result, incorrect use.

A.11 Functions for Enumerated Data Types

The standard functions SEL, MUX, EQ and NE can also be employed in some cases for enumerated data types. They are then applied as for integers (values of enumerations correspond to constants "coded" by the programming system).

B Standard Function Blocks

This appendix contains a complete overview of all the standard PLC function blocks described by means of examples in Chapter 2. For every standard function block of IEC 61131-3, the following information is given:

- Graphical declaration
- (Semantic) description
- Specification of some function blocks in Structured Text (ST).

In this book, the inputs and outputs of the standard FBs are given the names prescribed by the current version of IEC 61131-3. No allowance has been made for the fact that the possibility of calling FBs as "IL operators" can result in conflicts.

These conflicts occur with the IL operators (see also Section 4.1) LD, R and S, which need to be distinguished from the FB inputs LD, R and S when checking for correct program syntax. Depending on the programming system, these three formal operands could, for example, be called LOAD, RESET and SET in future.

B.1 Bistable Elements (Flip-Flops)

Figure B.1. Graphical declarations of the bistable function blocks SR and RS

```
FUNCTION_BLOCK     SR          (* flip-flop set dominant *)
VAR_INPUT
  S1   :   BOOL;
  R    :   BOOL;
END_VAR
VAR_OUTPUT
  Q1   :   BOOL;
END_VAR
Q1      :=  S1 OR ( NOT R AND Q1);
END_FUNCTION_BLOCK

FUNCTION_BLOCK     RS          (* flip flop reset dominant *)
VAR_INPUT
  S    :   BOOL;
  R1   :   BOOL;
END_VAR
VAR_OUTPUT
  Q1   :   BOOL;
END_VAR
Q1      :=  NOT R1 AND ( S OR Q1);
END_FUNCTION_BLOCK
```

Example B.1, Specification of the bistable function blocks SR and RS in ST

These two flip-flops implement dominant setting and resetting.

B.2 Edge Detection

Figure B.2. Graphical declarations of the function blocks R_TRIG and F_TRIG

```
FUNCTION_BLOCK      R_TRIG    (* rising edge *)
  VAR_INPUT
    CLK :    BOOL;
  END_VAR
  VAR_OUTPUT
    Q    :    BOOL;
  END_VAR
  VAR RETAIN                          (* example use of retentive variable *)
    MEM :    BOOL := 0;               (* initialise edge flag *)
  END_VAR
  Q      := CLK AND NOT MEM;          (* recognise rising edge *)
  MEM   := CLK;                       (* reset edge flag *)
END_FUNCTION_BLOCK

FUNCTION_BLOCK      F_TRIG    (* falling edge *)
  VAR_INPUT
    CLK :    BOOL;
  END_VAR
  VAR_OUTPUT
    Q    :    BOOL;
  END_VAR
  VAR RETAIN
    MEM :    BOOL := 1;               (* initialise edge flag *)
  END_VAR
  Q      := NOT CLK AND NOT MEM;      (* recognise falling edge *)
  MEM   := NOT CLK;                   (* reset edge flag *)
END_FUNCTION_BLOCK
```

Example B.2. Specification of the function blocks R_TRIG and F_TRIG in ST

In the case of the function blocks R_TRIG and F_TRIG, it should be noted that they detect an "edge" on the **first** call if the input of R_TRIG is TRUE or the input of F_TRIG is FALSE.

B.3 Counters

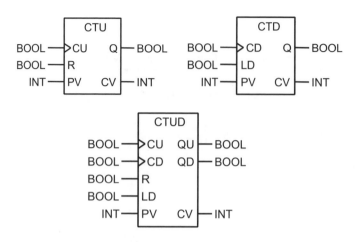

Figure B.3. Graphical declarations of the function blocks CTU, CTD and CTUD

```
FUNCTION_BLOCK        CTU      (* up counter *)
  VAR_INPUT
    CU  :   BOOL   R_EDGE;     (* CU with rising edge *)
    R   :   BOOL;
    PV  :   INT;
  END_VAR
  VAR_OUTPUT
    Q   :   BOOL;
    CV  :   INT;
  END_VAR
  IF  R  THEN                  (* reset counter *)
       CV    :=  0;
  ELSIF  CU AND ( CV < PV)  THEN
       CV    :=  CV + 1;       (* count up *)
  ENDIF;
  Q    :=    (CV >= PV);       (* limit reached *)
END_FUNCTION_BLOCK
```

Example B.3. Specification of the function blocks CTU and CTD in ST (continued on next page)

When used as versions for DINT, LINT, UDINT, ULINT, the counters CTU and CTD are accordingly designated CTU_DINT etc. up to CTD_ULINT.

```
FUNCTION_BLOCK        CTD        (* down counter *)
  VAR_INPUT
    CD   :   BOOL    R_EDGE;    (* CD with falling edge *)
    LD   :   BOOL;
    PV   :   INT;
  END_VAR
  VAR_OUTPUT
    Q    :   BOOL;
    CV   :   INT;
  END_VAR
  IF  LD  THEN                  (* reset counter *)
        CV      :=  PV;
  ELSIF  CD AND ( CV > 0)  THEN
        CV      :=  CV - 1;     (* count down *)
  ENDIF;
  Q    :=   (CV <= 0);          (* zero reached *)
END_FUNCTION_BLOCK
```

Example B.3. (Continued)


```
FUNCTION_BLOCK        CTUD       (* up-down counter *)
  VAR_INPUT
    CU   :   BOOL    R_EDGE;     (* CU with rising edge *)
    CD   :   BOOL    R_EDGE;     (* CD with falling edge *)
    R    :   BOOL;
    LD   :   BOOL;
    PV   :   INT;
  END_VAR
  VAR_OUTPUT
    QU   :   BOOL;
    QD   :   BOOL;
    CV   :   INT;
  END_VAR
  IF  R  THEN                    (* reset counter (reset dominant) *)
        CV   :=   0;
  ELSIF  LD  THEN
        CV   :=   PV;            (* set to count value *)
  ELSE
    IF NOT (CU AND CD) THEN
      IF  CU AND ( CV < PV)  THEN
        CV   :=   CV + 1;        (* count up *)
      ELSIF  CD AND ( CV > 0)  THEN
        CV   :=   CV - 1;        (* count down *)
      ENDIF;
    ENDIF;
  ENDIF;
  QU   :=   (CV >= PV);          (* limit reached *)
  QD   :=   (CV <= 0);           (* zero reached *)
END_FUNCTION_BLOCK
```

Example B.4. Specification of the function block CTUD in ST

B.4 Timers

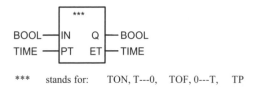

*** stands for: TON, T---0,   TOF, 0---T,   TP

Figure B.4. Graphical declarations of the function blocks TON, TOF and TP

The timers TP, TON and TOF are specified here using timing diagrams.

This time behaviour is only possible if the cycle time of the cyclic PLC program in which the timer is used is negligibly small in comparison with the duration PT if the timer is called only once in the cycle.

The diagrams show the behaviour of outputs Q and ET depending on input IN. The time axis runs from left to right and is labelled "t". The Boolean variables IN and Q change between "0" and "1" and the time value ET increases as shown.

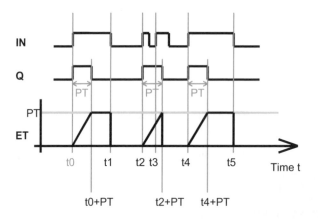

Figure B.5. Timing diagram for pulse timer TP depending on input IN

The standard FB "TP" acts as a pulse generator which supplies a pulse of constant length at output Q when a rising edge is detected at input IN. The time that has elapsed so far can be read off at output ET at any time.

As can be seen from Figure B.5, timers of type TP are not "retriggerable". If the intervals between the input pulses at IN are shorter than the pre-set time period, the pulse duration still remains constant (see period [t2; t2+PT]). Timing therefore does **not** begin again with every rising edge at IN.

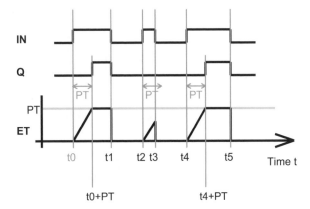

Figure B.6. Timing diagram for on-delay timer TON depending on input IN

The on-delay timer TON supplies the input value IN at Q with a time delay when a rising edge is detected at IN. If input IN is "1" only for a short pulse (shorter than PT), the timer is not started for this edge.

The elapsed time can be read off at output ET.

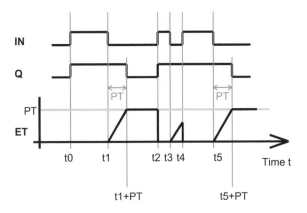

Figure B.7. Timing diagram for off-delay timer TOF depending on input IN

The off-delay timer performs the inverse function to TON i.e. it delays a falling edge in the same way as TON delays a rising one.

The behaviour of the timer TOF if PT is modified during timer operation is implementation-dependent.

C IL Examples

This appendix contains full examples of PLC programming with IEC 61131-3 for each type of POU, to supplement the information given in Chapters 2 and 4.

These examples are to be found on the CD enclosed in this book.

C.1 Example of a FUNCTION

The function ByteExtr extracts the upper or lower byte of an input word and returns it as the function value:

```
FUNCTION  ByteExtr : BYTE   (* extract byte from word *)
                            (* beginning of declaration part *)
VAR_INPUT                   (* input variables *)
  DbByte : WORD;            (* word consists of upper + lower byte *)
  Upper  : BOOL;            (* TRUE: take upper byte, else lower *)
END_VAR
                            (* beginning of instruction part *)
LD      Upper               (*extract upper or lower byte? *)
EQ      FALSE               (* lower? *)
JMPCN   UpByte              (* jump in the case of extraction of the upper byte *)
LD      DbByte              (* load word *)
WORD_TO_BYTE                (* conversion for type compatibility *)
ST      ByteExtr            (* assignment to the function value *)
RET                         (* nothing to do *)
UpByte:                     (* jump label *)
LD      DbByte              (* load word *)
SHR     8                   (* shift upper byte 8 bits to the right *)
WORD_TO_BYTE                (* conversion for type compatibility *)
ST      ByteExtr
RET                         (* return with function value in CR *)
                            (* end of FUN *)
END_FUNCTION
```

Example C.1. Example of the declaration of a function in IL

The ByteExtr function in Example C.1 has the input parameter DbByte of type WORD and the Boolean input Upper. The value returned by the function is of type BYTE. This function requires no local variables. The VAR ... VAR_END section is therefore missing from the declaration part.

The returned value is in the current result (CR) when the function returns to the calling routine with RET. At this point (jump label End:) the CR is of data type WORD, because DbByte was previously loaded. The function value is of data type BYTE after type conversion.

IEC 61131-3 always requires a strict "type compatibility" in cases like this. It is the job of the programming system to check this consistently. This is why a standard type conversion function (WORD_TO_BYTE) is called in Example C.1.

Example C.2 shows the instruction part of ByteExtr in the ST programming language.

```
FUNCTION  ByteExtr : BYTE      (* extraction byte from word *)
VAR_INPUT ... END_VAR          (* as above *)
IF Upper THEN
  ByteExtr := WORD_TO_BYTE (SHR (DbByte, 8) );
ELSE
  ByteExtr := WORD_TO_BYTE (DbByte);
END_IF;
END_FUNCTION
```

Example C.2. Instruction part of Example C.1 in ST

FUNCTION

END_FUNCTION

Example C.3. Graphical declaration part of the function declaration in Example C.1 (left) with an example of a call (right)

Example C.3 shows the declaration part and an example of a call for the function ByteExtr in graphical representation. The call replaces the lower byte of flag word %MW4 by its upper byte.

C.2 Example of a FUNCTION_BLOCK

The function block DivWithRem calculates the result of dividing two integers and returns both the division result and the remainder. "Division by zero" is indicated by an output flag.

```
FUNCTION_BLOCK DivWithRem    (* division with remainder *)
                             (* beginning of declaration part *)
VAR_INPUT                    (* input parameter *)
   Dividend :   INT;         (* integer to be divided *)
   Divisor   :   INT;         (* integral divisor *)
END_VAR
VAR_OUTPUT RETAIN            (* retentive output parameters *)
   Quotient :   INT;         (* result of the division *)
   DivRem   :   INT;         (* remainder after division *)
   DivError  :   BOOL;        (* flag for division by zero *)
END_VAR
                             (* beginning of instruction part *)
LD       0                   (* load zero *)
EQ       Divisor             (* divisor equal to zero? *)
JMPC     Error               (* catch error condition *)
LD       Dividend            (* load dividend, divisor not equal to zero *)
DIV      Divisor             (* carry out division *)
ST       Quotient            (* store integral division result *)
MUL      Divisor             (* multiply division result by divisor *)
ST       DivRem              (* store interim result *)
LD       Dividend            (* load dividend *)
SUB      DivRem              (* subtract interim result *)
ST       DivRem              (* yields "remainder" of the division as an integer*)
LD       FALSE               (* load logical "0" for error flag *)
ST       DivError            (* reset error flag *)
JMP      End                 (* ready, jump to end *)
Error:                       (* handling routine for error "division by zero" *)
LD       0                   (* zero, since outputs are invalid in event of error *)
ST       Quotient            (* reset Result *)
ST       DivRem              (* reset Remainder *)
LD       TRUE                (* load logical "1" for error flag *)
ST       DivError            (* set error flag *)
End:
RET
                             (* end of FB *)
END_FUNCTION_BLOCK
```

Example C.4. Example of the declaration of a function block in IL

The FB DivWithRem in Example C.4 performs integer division with remainder on the two input parameters Dividend and Divisor. In the case of division by zero, the error output DivError is set and the other two outputs are set in a defined manner to zero since they are invalid. The outputs are retentive, i.e. they are retained within the FB instance from which DivWithRem was called.

This example can also be formulated as a function because no statistical information has to be retained between calls.

Example C.5 shows the instruction part of DivWithRem in the ST programming language.

```
FUNCTION_BLOCK DivWithRem    (* division with remainder *)
VAR_INPUT ... END_VAR          (* as above *)
VAR_OUTPUT RETAIN ... END_VAR
IF Divisor = 0 THEN
  Quotient := 0;
  DivRem := 0;
  DivError := TRUE;
ELSE
  Quotient := Dividend / Divisor;
  DivRem := Dividend - (Quotient * Divisor);
  DivError := FALSE;
END_IF;
END_FUNCTION_BLOCK
```

Example C.5. Instruction part of Example C.4 in ST

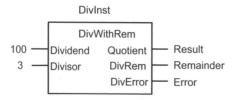

Example C.6. Graphical declaration part for Example C.4 (top) and example of a call of the FB instance DivInst (bottom)

Example C.6 shows the declaration part and an example of a call of the function DivWithRem in graphical representation. The FB must be instantiated (DivInst) before it can be called.

After execution of the function block, the output variables of this instance have the following values: DivInst.Quotient = 33, DivInst.DivRem = 1 and DivInst.DivError = FALSE. These return values are assigned to the variables Result, Remainder and Error respectively.

C.3 Example of a PROGRAM

The program MainProg in Example C.8 is not a complete programming example, but shows ways of implementing problems and illustrates the use of variables in POUs of type PROGRAM.

MainProg first starts a real-time clock DateTime that records the date and time by using function block RTC. This function block can be either user specific or predefined by manufacturer – since edition 2 of the standard IEC 1131-3 this RTC is not avaliable as a standard FB any longer. Example C.7 shows one possible representation as example. This real-time clock RTC is used to find out how long an interruption in the program has lasted (TimeDiff). The PLC system must be able to detect the interruption (Ress_Running) and must be able to access a hardware clock with an I/O address (ActDateTime).

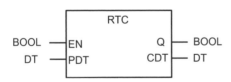

Example C.7. Graphical representation of example RTC, which is not a standard FB of IEC 1131-3. Rising edge at EN loads the real-time-clock, CDT is current date and time while EN; Q is copy of EN.

```
PROGRAM      MainProg              (* example of a main program *)
  VAR_INPUT                        (* input variables *)
    T_Start      : BOOL := FALSE;  (* input starting condition *)
  END_VAR
  VAR_OUTPUT                       (* output variables *)
    T_Failure    : BOOL := FALSE;  (* output "failure" *)
  END_VAR
  VAR_GLOBAL RETAIN                (* global retentive data area *)
    Ress_Running AT %MX255.5 : BOOL; (* running flag for resource/PLC-CPU *)
    DateTime     : RTC;            (* use buffered clock: date and time *)
    ActDateTime AT %MD2 : DT;      (* hardware clock: actual date with time *)
  END_VAR
  VAR_GLOBAL                       (* global data area *)
    EmergOff AT %IX255.0 : BOOL;   (* contact Emergency-Off *)
    ProgRun      : BOOL := FALSE;  (* "running" flag *)
    Error        : BOOL;           (* error flag *)
    Err_Code     : UDINT := 0;     (* Error code, 32 bit unsigned *)
  END_VAR
  VAR                              (* local variables *)
    AT %IX250.2 : BOOL;            (* directly represented variables *)
    ErrorProc    : CErrProc;       (* FB instance of CErrProc *)
    Edge         : R_TRIG;         (* edge detection *)
    TimeDiff     : TIME := t#0s;   (* time difference *)
  END_VAR
```

Example C.8. Example of the declaration of a main program in IL. The FB CErrProc ("central error processing") must already be available (continued on next page)

```
                                        (* beginning of instruction part *)
LD      FALSE
ST      Error                           (* reset error flag *)

...                             (* determine how long a power failure and/or an interruption *)
                                (* of the CPU lasted; Clock "DateTime" is battery-backed *)
LD      t#0s                            (* zero seconds *)
ST      TimeDiff                        (* reset *)
LD      DateTime.Q                      (* time valid = clock running *)
JMPC    Goon                            (* valid - nothing to do *)
LD      ActDateTime                     (* check current time *)
SUB     DateTime.CDT                    (* last time before power failure *)
ST      TimeDiff                        (* duration of power failure *)
Goon:
...
LD      Ress_Running                    (* CPU is running *)
ST      DateTime.IN                     (* clock running if CPU is running *)
LD      ActDateTime                     (* initial value Date/Time *)
ST      DateTime.PDT                    (* initial value of clock is loaded *)
                                        (* if there is a rising edge at IN *)
CAL     DateTime                        (* start real-time clock)
...
LD      TimeDiff                        (* interruption duration *)
LT      t#50m                           (* less than 50 seconds? *)
JMPC    Continue                        (* plant still warm enough ... *)
NOT
S       Error                           (* set error flag *)
LD      16#000300F2
ST      Err_Code                        (* store error cause *)
Continue:
LD      EmergOff                        (* "emergency off" pressed? *)
ST      Edge.CLK                        (* input edge detection *)
CAL     Edge                            (* compare input with edge flag *)
LDN     Edge.Q                          (* edge at EmergOff recognized? *)
AND     T_Start                         (* AND start flag set *)
ANDN    Error                           (* AND no new error *)
AND     ErrorProc.Ack                   (* AND error cause repaired *)
ST      ProgRun                         (* global "running" condition *)

....                            (* ...real instructions, calls of FUN/FBs... *)

LD      Error                           (* error occurred? *)
AND     %IX250.2
R       ProgRun                         (* reset global starting condition *)
LD      ProgRun
JMPCN   End                             (* in the case of error: start error handler *)
CALC    ErrorProc(Code := Err_Code)     (* FB with central error handling *)
LD      ErrorProc.Ack                   (* flag: error acknowledged *)
End:
LD      ProgRun                         (* program not running *)
ST      T_Failure                       (* set output parameter *)
RET                                     (* return and/or end *)
```

END_PROGRAM

Example C.8. (Continued)

Edge detection is also employed in this program in order to find out whether the Emergency Off button has been pressed.

In the global data area, the variable ProgRun is declared, which is available to all function blocks called under MainProg (as VAR_EXTERNAL). This is linked with the starting condition provided EmergOff has not been pressed.

FB instance ErrProc can handle errors with error code Err_Code. When the error has been corrected and acknowledged, the corresponding output is set to TRUE.

```
RESOURCE CentralUnit_1 ON CPU_001
  TASK Periodic (INTERVAL := time#13ms, PRIORITY := 1);
  PROGRAM Applic WITH Periodic : MainProg (   T_Start := %I250.0,
                                              T_Failure => %Q0.5);
END_RESOURCE
```

Example C.9. Resource definition with run-time program Applic for Example C.8

The program MainProg is provided with the properties of the periodic task Periodic and becomes the run-time program Applic of the resource (PLC-CPU) CentralUnit_1. This run-time program runs with highest priority (1) as a periodic task with a maximum cycle time of 13 ms.

   MainProg is called with the value of input bit %I250.0 for the input variable T_Start and sets output bit %Q0.5 with the value of T_Failure.

D Standard Data Types

This appendix summarises all the elementary data types and their features in tabular form. Their use is explained in Chapter 3.

IEC 61131-3 defines five groups of elementary data types. Their generic data types are indicated in brackets (see Table 3.9):

- Bit string   (ANY_BIT),
- Integer, signed and unsigned   (ANY_INT),
- Real (floating-point)   (ANY_REAL),
- Date, time-of-day   (ANY_DATE),
- Character string, duration   (ANYSTRING, TIME).

The following information is given for each data type, listed in separate tables for each group:

- Name (keyword),
- Description,
- Number of bits (data width),
- Value range (using the IEC literals),
- Initial values (default values).

The data width and permissible range of the data types in Tables D.5 and D.6 are application-dependent.

Data type	Description	Bits	Range	Initial
BOOL	Boolean	1	[0,1]	0
BYTE	Bit string 8	8	[0,...,16#FF]	0
WORD	Bit string 16	16	[0,...,16#FFFF]	0
DWORD	Bit string 32	32	[0,...,16#FFFF FFFF]	0
LWORD	Bit string 64	64	[0,...,16#FFFF FFFF FFFF FFFF]	0

Table D.1. "Boolean and Bit String" data types

Data type	Description	Bits	Range	Initial
SINT	Short integer	8	[-128,...,+127]	0
INT	Integer	16	[-32768,...,+32767]	0
DINT	Double integer	32	$[-2^{31},...,+2^{31}-1]$	0
LINT	Long integer	64	$[-2^{63},...,+2^{63}-1]$	0

Table D.2. . "Signed Integer" data types

Data type	Description	Bits	Range	Initial
USINT	Unsigned short integer	8	[0,...,+255]	0
UINT	Unsigned integer	16	[0,...,+65535]	0
UDINT	Unsigned double integer	32	$[0,...,+2^{32}-1]$	0
ULINT	Unsigned long integer	64	$[0,...,+2^{64}-1]$	0

Table D.3. "Unsigned Integer" data types

Data type	Description	Bits	Range	Initial
REAL	Real numbers	32	See IEC 559	0.0
LREAL	Long reals	64	See IEC 559	0.0

Table D.4. "Real Numbers" data types (floating-point numbers)

Data type	Description	initial
DATE	Date (only)	d#0001-01-01
TOD	Time of day (only)	tod#00:00:00
DT	Date and time of day	dt#0001-01-01-00:00:00

Table D.5. "Date and Time" data types

The full keyword TIME_OF_DAY can also be used in place of TOD, and DATE_AND_TIME in place of DT.

Data type	Description	Initial
TIME	Duration	t#0s
STRING	Character string (variable length)	''
WSTRING	Character string (double bytes)	''''''

Table D.6. "Duration and Character String" data types. The initial values for STRING and WSTRING are "empty" character strings.

E Causes of Error

IEC 61131-3 requires manufacturers to provide a list of responses to the following error conditions. See also Section 7.9. The responses fall into four different categories:

1) No system response (%),
2) Warning during program creation (WarnPc),
3) Error message during program creation (ErrPc),
4) Error message and response to error at run time (ErrRun).

No.	Error cause	Response
1	Nested comment	ErrPc
2	Ambiguous value of enumeration type	ErrPc
3	Value of a variable exceeds the specified range.	ErrRun
4	Incomplete address resolution in configuration ("*")	ErrPc
5	Write access to variable declared as CONSTANT	ErrPc
6	Variable declared as VAR_GLOBAL CONSTANT, but referenced as VAR_EXTERNAL without CONSTANT.	ErrPc
7	Reference to a directly represented or external variable in a function	ErrPc
8	Impermissible assignment to VAR_IN_OUT variable (e.g. to a constant)	ErrPc
9	Ambiguous assignment to VAR_IN_OUT variable	ErrPc
10	Type conversion errors	ErrPc
11	Numerical result (of a standard function) exceeds the range for the data type; Division by zero (in a standard function).	ErrRun ErrPc, ErrRun
12	Bit shift function with negative shift number	ErrRun
13	Mixed input data types to a selection function (standard function); Selector (K) out of range for MUX function.	ErrPc ErrRun

Table E.1. Error causes. The manufacturer supplies a table specifying the system response to the errors described above. The entry in the Response column marks the appropriate instant for the response in accordance with IEC 61131-3. (Continued on next page)

14	Character sequence functions: Invalid character position specified;	WarnPc, ErrRun
	Result exceeds maximum string length (INSERT/CONCAT result is too long).	WarnPc, ErrRun
	ANY_INT input is negative.	ErrRun
15	Result exceeds range for data type "Time"	ErrRun
16	An FB instance used as an input parameter has no parameter values.	ErrPc
17	A VAR_IN_OUT parameter has no value.	ErrPc
18	Zero or more than one initial steps in SFC network;	ErrPc
	User program attempts to modify step state or time.	ErrPc
19	Side effects in evaluation of transition conditions.	ErrPc
20	Action control contention error.	WarnPc, ErrRun
21	Simultaneously true, non-prioritised transitions in a SFC selection divergence.	WarnPc, ErrPc
22	Unsafe or unreachable SFC.	WarnPc
23	Data type conflict in VAR_ACCESS.	ErrPc
24	Tasks require too many processor resources;	ErrPc
	Execution deadline not met.	WarnPc, ErrRun
25	Numerical result exceeds range for data type (IL).	WarnPc, ErrRun
26	Current result (CR) and operand type do not match (IL).	ErrPc
27	Division by zero (ST);	WarnPc,ErrPc, ErrRun
	Invalid data type for operation (ST).	ErrPc
	Numerical result exceeds range for data type (ST).	ErrRun
28	Return from function without value assigned (ST).	ErrPc
29	Iteration fails to terminate (ST).	WarnPc,ErrPc, ErrRun
30	Same identifier used as connector label and element name (LD / FBD).	ErrPc
31	Uninitialised feedback variable	ErrPc

Table E.1. (Continued)

This table is intended as a guide. There are other possible errors which are not included in the IEC table. It should therefore be extended by every manufacturer as appropriate.

F Implementation-Dependent Parameters

Table F.1 lists the implementation-dependent parameters defined by IEC 61131-3. See also Section **7.10**.

No.	Implementation-dependent parameters
1	Maximum length of identifiers.
2	Maximum comment length (without leading or trailing brackets).
3	Pragma syntax and semantics.
4	Syntax and semantics for ", unless used to declare double-byte strings.
5	Range of values for variables of data type TIME (e.g. 0<=TimeVar<1000 Days), DATE, TIME_OF_DAY and DATE_AND_TIME. Precision of representation of seconds in data types TIME_OF_DAY and DATE_AND_TIME.
6	Maximum: - number of elements in an enumeration data type, - number of elements in an array (number of array subscripts), - array size (number of bytes), - number of structure elements, - structure size (number of bytes), - number of variables per declaration that can be declared with the same data type (separated by commas), - number of structure nesting depth.

Table F.1. Implementation-dependent parameters which every manufacturer of an IEC 61131-3 system must describe. The division into individual groups relates to individual sections in the standard. If no concrete figures are required, the manufacturer can add an informal description. (Continued on next page)

7	Default maximum length of STRING and WSTRING variables. Maximum allowed length of STRING and WSTRING variables.
8	Maximum number of hierarchical levels for directly represented variables (e.g. %QX1.1.1.2...). Logical or physical mapping (symbols %IX.... onto the real hardware).
9	Initialisation of system inputs (%IX...) at system start time.
10	Maximum number of variables per declaration block.
11	Behaviour of an AT identifier as part of an FB declaration.
12	Hot-restart behaviour, if variable was declared neither with RETAIN nor with NON_RETAIN.
13	Information to determine execution times of program organisation units (POU) (no details are given in IEC 61131-3).
14	Block output values, if block terminates with ENO=FALSE (yes/no).
15	Maximum number of function specifications (if limited).
16	Maximum number of inputs of extensible functions.
17	Effects of type conversions on accuracy (REAL_TO_INT, ...) and implemented error handling.
18	Accuracy of numerical functions.
19	Results of type conversions between TIME and other data types.
20	Maximum number of user function block specifications and instantiations (if limited).
21	Assignment behaviour of block inputs,if EN= FALSE (yes/ no).
22	Range of parameter PV (end value of counter function blocks).
23	System reaction to a change of the input PT (end value of timer function blocks) during operation.
24	Program size limitations (executable code).
25	Precision of step elapsed time; see Section 4.6.3
26	Maximum number of steps per SFC network.
27	Maximum number of transitions per SFC network and per step.
28	Maximum number of action blocks per step.
29	Action control mechanism (if available) and accessibility to Q and A output values.
30	(Minimum) transition clearing time (caused by PLC cycle time).
31	Maximum number of predecessor and successor steps in diverge/converge constructs.
32	Contents of RESOURCE libraries. Every processing unit (e.g. processor type) receives a description of all functions (standard functions, standard FBs, data types, ...) that can be processed by this resource.
33	Effect of READ_WRITE access on FB outputs

Table F.1. (Continued on next page)

34	Maximum number of tasks / resources. Task interval resolution (time between calls for periodic tasks); Type of task priority control (pre-emptive or non-pre-emptive scheduling).
35	Maximum length of expressions (ST) (number of operands, operators).
36	Maximum length of statements (ST) (IF; CASE; FOR; ...); restrictions.
37	Maximum number of CASE selections (ST).
38	Value of control variable upon termination of FOR loop.
39	Restrictions on network topology (LD/FBD).
40	Evaluation order of feedback loops.

Table F.1. (Continued)

There are a large number of other parameters that need to be considered when implementing a programming system compliant with IEC 61131-3.

Table F.2 gives a subjective selection of these parameters.

No.	Implementation-dependent parameters
1	Extent of syntax and semantic checking provided by the textual and graphical editors (during input).
2	Description of the execution order of networks.
3	Free placement of graphic elements in the case of editors with line updating ("elastic band") or automatic placement according to syntax rules.
4	Declaration and use of Directly Represented Variables (DRVs): - In the declaration the symbolic name is always used. In the code part *either* the symbol alone *or* both may be used (symbol; direct physical address). Or - DRVs may also be declared without symbolic names (use of the direct physical address only). Or - The programming system declares DRVs implicitly.
5	Sequence order of VAR_*...VAR_END blocks as well as multiple use of blocks with the same name (variable type).
6	Function calling mechanisms implemented in ST (with/without formal parameters).
7	Extent of implementation of the EN/ENO parameters in LD and FBD, and possible effects on IL/ST program sources.

Table F.2. Other implementation-dependent parameters (not part of IEC 61131-3). (Continued on next page)

8	Possibility of passing user-defined structures (TYPE) as function and FB input parameters (not defined in the standard at present).
9	Global and POU-wide publication of user-defined data types TYPE...END_TYPE: (not defined in the standard at present).
10	Extent of data area checking during program creation and at run time.
11	Graphical representation of qualifiers in variable declarations.
12	Restrictions on use of complex data types (function type also permitted for type "string" or user-defined types, ...?).
13	Algorithm for the evaluation of LD/FBD networks (see Sections 4.3.4 and **7**.3.1).
14	Aids for comprehensibility of cross-compiled programs (if implemented).
	...

Table F.2. (Continued)

IEC 61131-3 expressly allows functionality beyond the standard. However, they must be described. Table F.3 gives some examples.

No.	Extensions
1	Extending range checking to more data types than only integer (ANY_INT), e.g. ANY_NUM.
2	Accepting FB instances as array variables.
3	Permitting FB instances in structures.
4	Allowing overloading for user-defined functions, function blocks and programs.
5	Time when the step width is calculated in the case of FOR statements.
6	Possible use of pre-processor statements for literals, macros, conditional compiling, Include statements (for input of files with FB interface information/prototypes, with the list of directly represented variables or EXTERNAL declarations employed, ...).
7	Use of different memory models in the PLC (Small; Compact; Large,...).
	...

Table F.3. Possible extensions (not part of IEC 61131-3)

G IL Syntax Example

Many of the examples given in this book are formulated in Instruction List (IL). This programming language is widely used and is supported by most programming systems. By including data types previously only found in high-level languages, such as arrays or structures, the IEC 61131-3 Instruction List language opens up new possibilities compared with conventional IL.

The IL syntax descriptions in this appendix are presented in simplified form in *syntax diagrams*.

Syntax diagrams explain the permissible use of delimiters, keywords, literals and names in a readily comprehensible format. They can easily be put into textual form for the development of compilers and define the formal structure of a programming language *(syntax* of a language). The reader can use the diagrams for reference.

The syntax descriptions in this appendix go beyond IEC 61131-3 because, in addition to the pure syntax definitions, they also include semantic conditions (consistent use of the current result, use of function parameters, etc.). IEC 61131-3 only offers an informal description of this.

The rules are outlined in Section G.1. The use of the diagrams is explained in Section G.2 by means of an example.

G.1 Syntax Diagrams for IL

If a node in the syntax diagram has further subdivisions (sub-diagram), its name appears in italics. Keywords or terminal symbols which are not further subdivided appear in standard type. For clarity the use of comments is not included. As of version 2 of the standard, a comment is permissible wherever blanks (except for character strings) can be placed. A comment begins with (*, ends with *) and contains any number of alphanumeric characters in between **without** EOL. Comments cannot be nested.

IL instruction sequence:

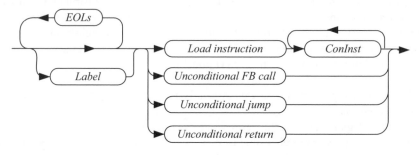

ConInst:

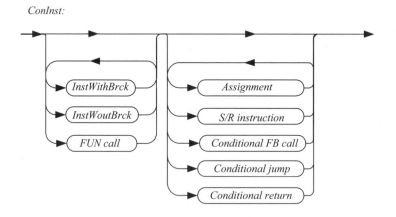

Figure G.1. Syntax diagrams of an IL instruction sequence with the sub-elements "conditional instruction" and "label" (continued on next page)

Label:

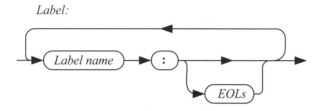

Figure G.1. (Continued)

An IL instruction sequence begins optionally with a (jump) label. This is followed either by a Load instruction followed by one or more conditional instructions ConInst, an unconditional instruction with FB call, a jump or a return (see syntax diagram in Figure G.1).

The conditional instruction begins with a sequence of instructions with and without brackets and/or function calls. This is followed by a series of (S/R) assignments, conditional calls or jumps.

The label consists of a label identifier, followed by a colon. It either immediately precedes the first instruction of the sequence or is followed by EOLs (end of line).

The syntax diagrams for instructions, calls and jumps are given below.

InstWoutBrck

Figure G.2. Syntax diagram of an instruction without brackets

An instruction without brackets (Figure G.2) consists of an IL operator with one operand and the end-of-line EOLs.

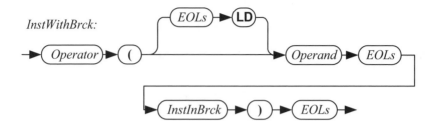

InstWithBrck:

InstInBrck:

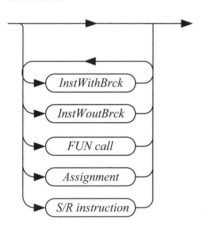

Figure G.3. Syntax diagrams of an instruction with brackets

The syntax diagrams in Figure G.3 show the structure of an instruction with brackets. This type of instruction begins with an instruction consisting of an opera-tor followed by an opening bracket and an operand. This can be followed by any number of instructions InstInBck inside the brackets, which are concluded with a closing bracket and EOLs.

These inner instructions can themselves contain brackets (nesting), as well as FUN calls and assignments.

FUN call with actual parameter:

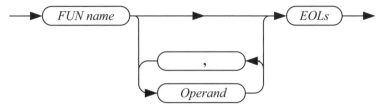

FUN call with formal parameter:

Figure G.4. Syntax diagram for a function call

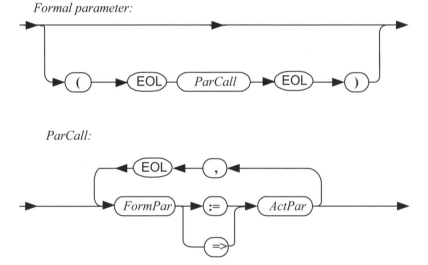

Figure G.5. Formal parameter assignment

As Figure G.4 shows, a function call consists of the function name together with a number of operands separated by commas as actual parameters of the function. Like a function block, a function can also be called with formal parameters.

FB call with actual parameter:

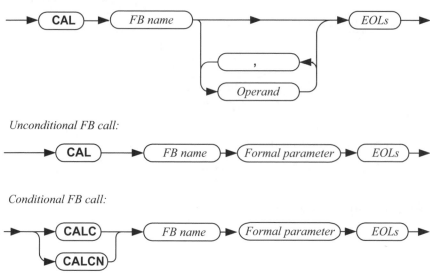

Unconditional FB call:

Conditional FB call:

Figure G.6. Syntax diagrams for a function block call

Figure G.6 shows the syntax diagrams for conditional and unconditional calls of a function block. The unconditional call begins with CAL, the conditional call with CALC or CALCN. This is followed by the name of the FB instance, and the FB parameters in brackets.

The assignment of an actual parameter to a formal parameter is represented by the symbol ":=". Such assignments are required for every parameter and are separated by commas.

FB call using Method 3

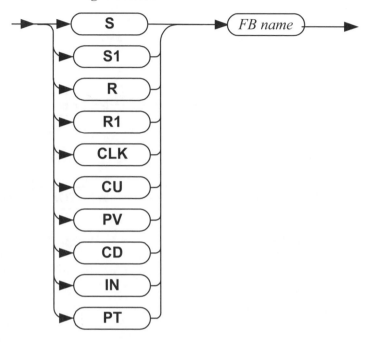

Figure G.7. FB call using Method 3

A call with the syntax shown in Figure G.7 results only in parameter initialisation or FB execution, depending on the parameters employed; see Section 4.12.

Unconditional jump:

Conditional jump:

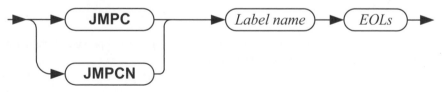

Figure G.8. Syntax diagrams for conditional and unconditional jumps

For jumps, the label name is specified after the jump operator JMP (unconditional) or JMPC/JMPCN (conditional) (Figure G.8).

Unconditional return:

Conditional return:

Figure G.9. Syntax diagrams for conditional and unconditional return

The returns shown in Figure G.9 have no operands or parameters.

Load instruction:

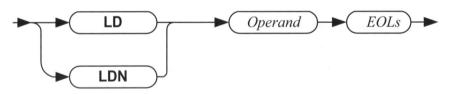

Figure G.10. Syntax diagram for the Load instruction

The Load instruction in Figure G.10 has a single (negatable) operand. It cannot be combined with a bracket or used inside a bracket.

Assignment

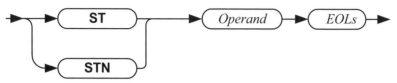

Figure G.11. Syntax diagram for assignment

Assignments (Figure G.11) consist of the operator ST or STN and the specification of the operand to be stored.

S/R instruction:

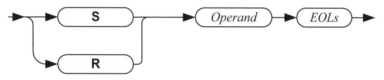

Figure G.12. Syntax diagram for the S/R instruction

An S/R instruction (Figure G.12) consists of the IL operators S or R and one operand.

Operator:

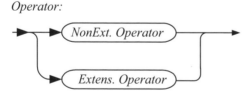

Figure G.13. Extensible and non-extensible operators

The operators represented in Figure G.13 perform logic operations, and are not used for loading or storage. A distinction is made between extensible and non-extensible operators, as shown below.

Extens. Operator:

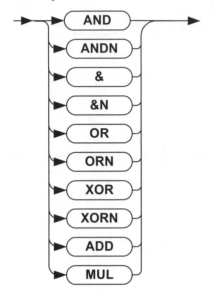

Figure G.14. Extensible operators: bitwise Boolean operations, addition and multiplication

Figure G.14 shows the extensible operators. They can have more than two input parameters. The bitwise Boolean operators (standard functions) can also be used with inversion.

NonExt. Operator:

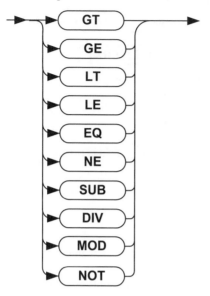

Figure G.15. Non-extensible operators: comparison, subtraction, division

The non-extensible operators in Figure G.15 have exactly two input parameters (including the current result CR).

EOLs:

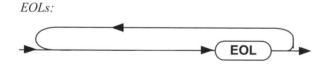

Figure G.16. Syntax diagram for the EOL (end of line) of an IL instruction

An IL line is concluded with a single EOL character (e.g. carriage return / line feed) or a comment followed by EOL (Figure G.16). These elements can occur once or any number of times in sequence.

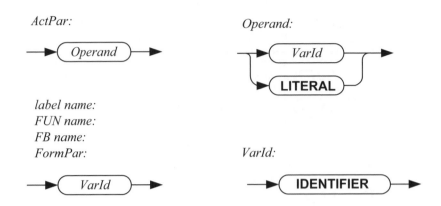

Figure G.17. Operands, parameters and other elements are represented by identifiers and literals.

Figure G.17 shows how parameters, operands and other elements are represented using IDENTIFIERs and LITERALs.

For simplicity the syntax diagrams of identifiers and literals are not shown here. The basic principles of their representation are explained in Section 3.2.

G.2 IL Example from Syntax Diagrams

The IL syntax diagrams shown on the previous pages will now be used to produce an IL example. This shows how sample programs are constructed from syntax diagrams and vice versa, enabling IL examples to be checked for correctness.

			Beginning of IL sequence
0001	SequenceOne:		(* label *)
0002			(* simple logic operation with jump *)
0003	LD	Var1	(* load instruction *)
0004			(* beginning of conditional instructions *)
0005	ANDN	Var2	(* instruction without bracket *)
0006	ORN	(Var3	(* instruction with bracket *)
0007	AND	Var4	
0008	)		(* end of bracketing *)
0009	AND	Var5	
0010	ST	Var6	(* assignment *)
0011	S	Var7	(* S/R instruction *)
0012	RETC		(* conditional return *)
0013			(* end of conditional instructions *)
0014			(* end of IL sequence *)

Example G.1. IL example. The comments refer to the corresponding syntax diagram. The line numbers on the left are used for reference in Table G.1.

To show how the IL example in Example G.1 can be built up from the syntax diagrams in Section G.1, Table G.1 shows the relevant syntax diagrams for each IL line.

Line in Ex. G.1	Syntax diagram	Figures
0001-0014	IL instruction sequence	Figure G.1
0001-0002	(Jump) label	Figure G.1
0003-0004	Load instruction	Figure G.10
0005	Instruction without bracket	Figure G.2
0005	Extensible operator ANDN	Figure G.14
0006-0008	Instruction with bracket	Figure G.3
0006,0007	Extensible operators ORN, AND	Figure G.14
0007,0009	Instruction without bracket	Figure G.2
0007,0009	Extensible operator AND	Figure G.14
0010	Assignment	Figure G.11
0011	S/R instruction	Figure G.12
0012-0014	Conditional return	Figure G.9

Table G.1. Syntax diagrams to be used for each line of the IL sequence in Example G.1

This example shows how a concrete IL program is built up using syntax diagrams. In the syntax diagram for an IL instruction sequence (Figure G.1), first the label with name, colon and comments is inserted, followed by the first (Load) instruction.

The "conditional instructions" part is made up of two instructions, the first of which with a bracket containing further instructions. After the conditional instructions, the sequence is terminated with assignments and return.

In this way it is possible to create valid IL sequences from the individual syntax diagrams. Conversely, the relevant syntax diagrams for each IL line can be found in order to determine whether a program section is syntactically correct.

H Reserved Keywords and Delimiters

IEC 61131-3 expressly permits the use of translation tables for adapting keywords and delimiters to national character sets.

H.1 Reserved Keywords

Table H.1 lists all reserved keywords for programming languages of IEC 61131-3 in alphabetical order. They must **not** be employed as names for user-defined elements. In order to prevent confusion, the table also includes all formal parameters for functions and/or FB as well as data types beyond the scope of the standard.

A	ABS	ACOS
	ACTION	ADD
	AND	ANDN
	ANY	ANY_BIT
	ANY_DATE	ANY_DERIVED
	ANY_ELEMENTARY	ANY_INT
	ANY_MAGNITUDE	ANY_NUM
	ANY_REAL	ARRAY
	ASIN	AT
	ATAN	
B	BOOL	BY
	BYTE	

Table H.1. Reserved keywords of IEC 61131-3 (continued on next page)

C	CAL	CALC
	CALCN	CASE
	CD	CDT
	CLK	CONCAT
	CONFIGURATION	CONSTANT
	COS	CTD
	CTU	CTUD
	CU	CV
D	D	DATE
	DATE_AND_TIME	DELETE
	DINT	DIV
	DO	DS
	DT	DWORD
E	ELSE	ELSIF
	END_ACTION	END_CASE
	END_CONFIGURATION	END_FOR
	END_FUNCTION	END_FUNCTION_BLOCK
	END_IF	END_PROGRAM
	END_REPEAT	END_RESOURCE
	END_STEP	END_STRUCT
	END_TRANSITION	END_TYPE
	END_VAR	END_WHILE
	EN	ENO
	EQ	ET
	EXIT	EXP
	EXPT	
F	FALSE	F_EDGE
	F_TRIG	FIND
	FOR	FROM
	FUNCTION	FUNCTION_BLOCK
G	GE	GT
I	IF	IN
	INITIAL_STEP	INSERT
	INT	INTERVAL

Table H.1. (Continued on next page)

J	JMP	JMPC
	JMPCN	
L	L	LD
	LDN	LE
	LEFT	LEN
	LIMIT	LINT
	LN	LOG
	LREAL	LT
	LWORD	
M	MAX	MID
	MIN	MOD
	MOVE	MUL
	MUX	
N	N	NE
	NEG	NON_RETAIN
	NOT	
O	OF	ON
	OR	ORN
P	P	PRIORITY
	PROGRAM	PT
	PV	
Q	Q	Q1
	QU	QD
R	R	R1
	R_EDGE	R_TRIG
	READ_ONLY	READ_WRITE
	REAL	RELEASE
	REPEAT	REPLACE
	RESOURCE	RET
	RETAIN	RETC
	RETCN	RETURN
	RIGHT	ROL
	ROR	RS

Table H.1. (Continued on next page)

S	S	S1
	SD	SEL
	SEMA	SHL
	SHR	SIN
	SINGLE	SINT
	SL	SQRT
	SR	ST
	STEP	STN
	STRING	STRUCT
	SUB	
T	T	TAN
	TASK	THEN
	TIME	TIME_OF_DAY
	TO	TOD
	TOF	TON
	TP	TRANSITION
	TRUE	TYPE
U	UDINT	UINT
	ULINT	UNTIL
	USINT	VAR
V	VAR_ACCESS	VAR_CONFIG .
	VAR_EXTERNAL	VAR_GLOBAL
	VAR_INPUT	VAR_IN_OUT
	VAR_OUTPUT	VAR_TEMP
W	WHILE	WITH
	WORD	WSTRING
X	XOR	XORN

Table H.1. (Continued)

H.2 Delimiters

Delimiters are "symbols" in the syntax of programming languages and have different meanings depending on where they are used. For example, round brackets can be used to indicate the beginning and end of a list of actual parameters in a function call, or they can be used together with the asterisk to frame comments.

All the delimiters and their combinations are listed in Table H.2 together with their possible meanings.

Characters for the graphical representation of lines are not included here.

Delimiter		Meaning, explanations
Space		Can be inserted anywhere - except within keywords, literals, identifiers, directly represented variables or combinations of delimiters (such as "(*" or "*)"). IEC 61131-3 does not specify any rules about tabulators (TABs). They are usually treated as spaces.
End of line (EOL)		Permissible at the end of a line in IL. In ST also permissible within statements. Not permitted in IL comments. EOL (end of line) is normally implemented by CR&LF (Carriage Return & Line Feed).
Begin comment	(*	Beginning of a comment (nesting not allowed)
End comment	*)	End of a comment
Plus	+	1. Leading sign of a decimal literal, also in the exponent of a real (floating-point) literal 2. Addition operator in expressions
Minus	-	1. Leading sign of a decimal literal, also in the exponent of a real (floating-point) literal 2. Subtraction operator in expressions 3. Negation operator in expressions 4. Year-month-day separator in time literals
Number sign ("hash")	#	1. Based number separator in literals 2. Time literal separator

Table H.2. Delimiters of IEC 61131-3 (continued on next page)

Delimiter	Meaning, explanations
Point .	1. Integer/fraction separator 2. Separator in the hierarchical addresses of directly represented and symbolic variables 3. Separator between components of a data structure (for access) 4. Separator for components of an FB instance (for access)
e, E	Leading character for exponents of real (floating-point) literals
Quotation mark '	Beginning and end of character strings
Dollar sign $	Beginning of special characters within character strings
Prefix time literals t#, T# d#, D# d, D h, H m, M s, S ms, MS date#, DATE# time#, TIME# time_of_day# TIME_OF_DAY# tod#, TOD# date_and_time# DATE_AND_TIME# dt#, DT#	Characters introducing time literals. Combinations of lower-case and upper-case letters are also permissible.
Colon :	Separator for: 1. Time within time literals 2. Data type specification in variable declarations 3. Data type name specification 4. Step names 5. PROGRAM...WITH... 6. Function name/data type 7. Access path: Name/type 8. Jump label before next statement 9. Network label before next statement
Assignment (1) :=	1. Operator for initial value assignment 2. Input connection operator (assignment of actual parameter to formal parameter in POU-call) 3. Assignment operator
Assignment (2) =>	Output connection operator (assignment of formal parameters to actual parameters in a PROGRAM call)

Table H.2. (Continued on next page)

Delimiter	Meaning, explanations
Round brackets (...)	Beginning and end of: 1. Enumeration list 2. Initial value list, also: multiple initial values (with repetition number) 3. Range specification 4. Array subscript 5. Character string length 6. Operator in IL (computation level) 7. Parameter list in POU call 8. Sub-expression hierarchy
Square brackets [...]	Beginning and end of: 1. Array subscript (access to an array element) 2. Character string length (in declaration)
Comma ,	Separator for: 1. Enumeration list 2. Initial value list 3. Array subscripts (multidimensional) 4. Variable names (in the case of multiple declarations with the same data type) 5. Parameter list in POU call 6. Operand list in IL 7. CASE value list
Semicolon ;	End of: 1. Definition of a (data) type 2. Declaration (e.g. variables) 3. ST statement
Two points ..	Separator for: 1. Range specification 2. CASE range
Percent %	Leading character for hierarchical addresses of directly represented and symbolic variables
Comparison >, < >=, <=, =, <>,	Relational operators in expressions
Exponent **	Operator in expressions
Multiplication *	Multiplication operator in expressions
Division /	Division operator in expressions
Ampersand &	AND operator in expressions

Table H.2. (Continued)

I Glossary

In this chapter important terms and abbreviations employed in the book are listed in alphabetical order and explained in detail.

Terms, which are defined by IEC 61131-3, are marked "IEC".

Action	IEC	Boolean variable or a series of statements which can be accessed via an *action block* (in *SFC*).
Action block	IEC	Activation description of *actions* (in *SFC*) using an associated control structure.
Action control	IEC	Control unit for every *action* in SFC which is supplied with the input condition for activating the assigned action by means of one or more *action blocks*; also: action control block.
Actual parameter		Actual value for an input variable (*formal parameter*) of a *POU*.
Allocation list		List which contains the assignment of all symbols or *symbolic variables* to *PLC* addresses.
Array		Sequence of elements of the same *data type*.
Block		Programming unit, from which PLC-programs are built. Blocks can often be loaded independently from each other into the PLC, see also *POU*. Also referred to as sub-program in other programming languages.
Cold restart	IEC	Restart of the PLC-system and its application program, whereby all variables and memory areas (such as internal registers, timers, counters) are (newly) initialised with predefined values. This process can occur automatically after specific events (e.g. after a power failure) or also manually by the user (e.g. Reset button).

Comment	IEC	Text written between parentheses and asterisks used to explain the program (cannot be nested!). This is not interpreted by the *programming system.*
Configuration	IEC	Language element CONFIGURATION which corresponds to a *PLC system*
CPU		Abbreviation for Central Processing Unit (e.g. of a PLC)
CR		Abbreviation for *Current Result*
Cross-compilation		Conversion of the representation of a *POU* from one programming language to another, typically between ladder and function block diagram, but also between textual and graphical languages; also: cross-compiling
Current result	IEC	Interim result in *IL* of any *data type*
Cycle		A single run of the (periodically called) application program.
Cycle time		The time which an application program requires for one *cycle*; also: scan time.
Data block		Shared data area, which is accessible throughout a program, see also *block.* In IEC 61131-3, there is no direct analogy. They are replaced here by global (non-local), structured data areas and *FB instance* data areas.
Data type		Defines properties of a *variable* as bit length and value range.
Declaration		Definition of *variables* and *FB instances* takes place in a *declaration block* with information about the data name, the *data type* or the *FB type* as well as appropriate *initial values*, *range specification* and *array attributes* (data template declaration).
		The definition or programming of *POUs* is also designated as a *declaration* since their interfaces are made known to the *programming system* here.
Declaration block	IEC	Combination of *declarations* of one variable type at the beginning of the *POU.*
Derived data type	IEC	With the aid of a *type definition,* a user-specific *data type* is created. Its elements are *elementary data types* or *derived data types.*
Directly represented variable	IEC	*Variable* without further name which corresponds to a *hierarchical address.*
Edge		The 0→1 transition of a Boolean variable is known as "rising edge". Accordingly, the 1→0 transition is known as "falling" edge.

Elementary data type	IEC	A standard *data type* predefined by IEC 61131-3.
Enumeration		Special *data type* for the definition of a range of textual sequences. Mostly implemented internally as integer values.
Extension of functions	IEC	A *function* can have a variable number of inputs.
FB	IEC	Abbreviation for *function block*
FB instance	IEC	see *instance*
FB type	IEC	Name of a *function block* with call and return interface
FBD	IEC	Abbreviation for *Function Block Diagram*
Formal parameter		Name or placeholder of an input variable (all *POUs*) or output variable (*function block* and *program*).
Function	IEC	A *POU* of type FUNCTION
Function block	IEC	A *POU* of type FUNCTION_BLOCK
Function Block Diagram	IEC	Function Block Diagram (FBD) is a programming language used to describe networks with Boolean, arithmetic and similar elements, working together concurrently.
Generic data type	IEC	Combination of *elementary data types* into groups using the prefix 'ANY', in order to describe *overloaded functions*.
Hierarchical address	IEC	Physical slot address of I/O modules of a *PLC system* (see also *I/O*).
Hot restart	IEC	Program restart at the place in the program where a power failure occurred. All battery-backed data areas as well as the application program context will be restored and the program can go on running, as if there had been no power failure. In contrast to *warm restart,* the interruption duration must be within a given value range depending on the process. For this purpose, the PLC system must have a separately secured real-time clock in order to be able to compute the interruption duration.
I/O		The addresses of input and output modules belonging to a *PLC system* with *hierarchical addresses*.
IL	IEC	Abbreviation for *Instruction List*
Indirect FB call		Call of an *FB instance* whose name is passed to the *POU* as a VAR_IN_OUT parameter.
Initial value		Value of a *variable*, which will be assigned during initialisation, i.e. at system start-up time; also: starting count.

Instance	IEC	Structured data set of an *FB* obtained by *declaration* of a *function block* indicating the *FB type*.
Instruction List	IEC	Instruction List (IL) is a much used Assembler-like programming language for PLC systems.
Ladder Diagram	IEC	Ladder Diagram (LD) is a programming language to describe networks with Boolean and electromechanical elements, such as contacts and coils, working together concurrently.
LD	IEC	Abbreviation for *Ladder Diagram*
Multi-element variable	IEC	*Variable* of type *array* or *structure*, which is put together from several different *data types*.
New start		see *Cold restart*
Overloading of functions	IEC	The capability of an operation or *function* to operate with different input *data types* (but each of the same type). By this means several function classes are available under the same name.
PC		Abbreviation for personal computer.
	IEC	Also Abbreviation for Programmable Controllers as employed in IEC 61131
PLC		Abbreviation for Programmable Logic Controller
PLC programming computer		Unit consisting of computer, *programming system* and other peripherals for programming the *PLC*.
PLC programming system		Set of programs which are necessary for programming a *PLC system:* program creation and compilation, transfer into the *PLC* as well as program test and commissioning functions.
PLC system		All hardware, firmware and operating system parts required for executing a PLC program.
POU	IEC	Abbreviation for *program organisation unit*
Pragma		Pragmas are offered by the *programming system* and are typically employed for program pre-processing and post-processing.
Program	IEC	A *POU* of type PROGRAM
Program organisation unit	IEC	A *block* of type *function, function block* or *program*, from which application programs are built.
Programming computer		see *PLC programming computer*
Programming system		see *PLC programming system*
Range specification		Specification of a permissible range of values for a *data type* or a *variable*.

Recursion		Illegal in IEC 61131-3. It means:
		a) the *declaration* of an *FB* using its own name or *FB type*,
		b) mutual *FB* calls.
		Recursion is considered an error and must be recognised while programming and/or at run time.
Resource	IEC	Language element RESOURCE which corresponds to a central processing unit of the *PLC system*.
Retentive data	IEC	Ability of a PLC to protect specific process data against loss during a power failure. The keyword RETAIN is used for this in IEC 61131-3.
Reverse compiling		Recovery of the *POU* source back from the *PLC* memory.
Run-time program		Program of *POU type* PROGRAM as an executable unit (by association with a *task*).
Semantics		Meaning of language elements of a programming language as well as their description and their application.
Sequential Function Chart	IEC	Sequential Function Chart (SFC) is a programming language used to describe sequential and parallel control sequences with time and event control.
SFC	IEC	Abbreviation for *Sequential Function Chart*
Single-element variable	IEC	*Variable* which is based on a single *data type*.
ST	IEC	Abbreviation for *Structured Text*
Standard function blocks	IEC	Set of *function blocks* predefined by IEC 61131-3 for implementation of typical PLC requirements.
Standard functions	IEC	Set of *functions* predefined by IEC 61131-3 for the implementation of typical PLC requirements.
Std. FB		Abbreviation for *standard function block*
Std. FUN		Abbreviation for *standard function*
Step	IEC	State element of an SFC program in which statements of the *action* corresponding to this *step* can be started.
Structured Text	IEC	Structured Text is a programming language used to describe algorithms and control tasks by means of a modern high programming language similar to PASCAL.
Symbolic variable	IEC	*Variable* with name (identifier) to which a *hierarchical address* is assigned.
Syntax		Structure and interaction of elements of a programming language.

Task	IEC	Definition of run-time properties of programs.
Transition	IEC	Element of an SFC program for movement from one SFC *step* to the next by evaluating the transition condition.
Type definition		Definition of a user-specific *data type* based on already available *data types*.
Variable		Name of a data memory area which accepts values defined by the corresponding *data type* and by information in the variable *declaration*.
Warm reboot		Term used for either *hot restart* or *warm restart*
Warm restart (at the beginning of the program)	IEC	Program restart similar to *hot restart* with the difference that the program starts again at the beginning, if the interruption duration exceeded the maximum time period allowed. The user program can recognise this situation by means of a corresponding status flag and can accordingly preset specific data.

J Bibliography

Books concerning PLC programming according to IEC 61131-3:

[JohnTiegel-09]   K.-H. John and M. Tiegelkamp
                   „SPS-Programmierung mit IEC 61131-3". Springer Berlin
                   Heidelberg New York, 2009, 4th Ed. (German),
                   ISBN 3-540-67752-6

[Lewis-98]   R. W. Lewis
                   „Programming industrial control systems using
                   IEC 1131-3", IEE Control Engineering, The Institution of
                   Electrical Engineers, 1998,
                   ISBN 0-852-96950-3

[Lewis-01]   R. W. Lewis
                   „Modelling Control Systems Using Iec 61499: Applying
                   Function Blocks to Distributed Systems ", IEE Control
                   Engineering, The Institution of Electrical Engineers, 2001,
                   ISBN 0-852-96796-9

[Bonfatti-03]   Dr. Monari, Prof. Bonfatti and Dr. Sampieri
                   „IEC 61131-3 Programming Methodology; Software
                   engineering methods for industrial automated systems",
                   2003, ISBN 978-0973467000

Standards concerning PLC programming:

[IEC 61131-1]   IEC 61131-1 Edition 2.0 (2003-05), TC/SC 65B.
                    Programmable controllers - Part 1: General information
                    www.iec.ch

[IEC 61131-2]   IEC 61131-2 Edition 3.0 (2007-07), TC/SC 65B,
                    Programmable controllers - Part 2: Equipment requirements
                    and tests
                    www.iec.ch

[IEC 61131-3]   IEC 61131-3 Edition 2.0 (2003-01), TC/SC 65B,
                    Programmable controllers - Part 3: Programming languages
                    www.iec.ch

[IEC 61131-4]   IEC 61131-4 Edition 2.0 (2004-07), TC/SC 65B,
                    Programmable controllers - Part 4: User guidelines
                    www.iec.ch

[IEC 61131-5]   IEC 61131-5 Edition 1.0 (2000-11), TC/SC 65B,
                    Programmable controllers - Part 5: Communications
                    www.iec.ch

[IEC 61131-7]   IEC 61131-7 Edition 1.0 (2000-08), TC/SC 65B,
                    Programmable controllers - Part 7: Fuzzy control
                    programming
                    www.iec.ch

[IEC 61131-8]   IEC 61131-8 Edition 2.0 (2003-09), TC/SC 65B,
                    Programmable controllers - Part 8: Guidelines for the
                    application and implementation of programming languages
                    www.iec.ch

[IEC 61499-1]   IEC 61499-1 Edition 1.0 (2005-01), TC/SC 65B,
                    Function blocks - Part 1: Architecture
                    www.iec.ch

[IEC 61499-2]   IEC 61499-2 Edition 1.0 (2005-01), TC/SC 65B,
                    Function blocks - Part 2: Software tool requirements
                    www.iec.ch

[IEC 61499-4]   IEC 61499-4 Edition 1.0 (2005-08), TC/SC 65B,
                    Function blocks - Part 4: Rules for compliance profiles
                    www.iec.ch

Other important references

[Holobloc]   HOLOBLOC, Inc., FBDK (Function Block Development Kit) technology for 61499
http://www.holobloc.com/

[IEC 65B WG7]   Working group for IEC 61131: „SC 65B-WG 7: Programmable control systems for discontinuous industrial-processes",
http://www.iec.ch/dyn/www/f?p=102:14:0::::FSP_ORG_ID:2598

[OPC Foundation] OPC Foundation, dedicated to ensuring interoperability in automation (formerly: OLE for Process Control)
http://www.opcfoundation.org

[PLCopen Europe] Eelco van der Wal
Molenstraat 34
4201 CX Gorinchem, The Netherlands

Tel: +31-183-660261
Fax: +31-183-664821
Email: evdwal@plcopen.org
http://ww.plcopen.org

[PLCopen North America]
Bill Lydon
10308 W. Cascade Dr.
Franklin, WI 53132
USA

Tel: +1 414-427-5853
Email: blydon@plcopen-na.org
http://www.plcopen-na.org/

[PLCopen Japan]   Shigeo Kawashima
Mitsui Sumitomo Bank Ningyo-cho
5-7, Nihonbashi Ohdemma-cho, Chuo-lu
Tokyo, 103-0011
Japan

Tel: +81-5847-8058
Fax: +81-5847-8180
Email: kawashima-shigeo@fujielectric.co.jp
http://www.plcopen-japan.jp

[PLCopen China]   No. 1, JiaoChangKuo
DeShengMenWai
Beijing, 100011
China

Tel: +86-(010) 6202-9216
Fax: +86-(010) 6202-7873
http://www.plcopen.org.cn

K Index